Hans J. Mayland
Dieter Bork

South American Dwarf Cichlids

Apistogramma
Crenicara
Microgeophagus

Verlag: A.C.S. GmbH, Germany

To the reader's attention

All the informations given in this book have been provided by the author to the best of his knowledge and judgement and were examined and reconsidered by author and publisher with greatest possible carefulness. If there should still be any mistakes author and publisher will not take responsibility for those and will not accept any claim for damages. This exclusion of liability does not concern the "Produkthaftung" after German law. Each reader has to be cautious when dealing with the substances, material, equipment etc. named in this book, he has to follow the instructions given by the producers of them and has to prevent that unauthorized persons deal with them.

Acknowledgement

We would like to thank all the persons and institutions which gave us help and advice in preparing this book.
We are particularly grateful

Institut Fresenius
Chemische und Biologische Laboratorien GmbH

to the
"Institut Fresenius"
for the water analysis
made by them free of charge

and to

VARIG Brasil

for their excellent cooperation as to the many flights in Brazil.

Photos:

Stefan Bergleiter, p. 13
Jürgen Elsässer, p. 60, 61
Jürgen Glaser, p. 55, 78, 165
Horst Linke, p. 139
Hans Joachim Richter, p. 77, 167
Frans Vermeulen, p. 173 (2)
Frank Warzel, p. 122, 171 (2)
Uwe Werner, p. 131 (2)
Wolfgang Windisch, p. 138 (below)
all the other photos are made by the authors

Coloured maps:

Richard Achilles, p. 10/11, 14/15

Black and white maps and drawings:

Hans J. Mayland

Translation from German language: Ishla Eberl, Schorndorf

Production: Landbuch-Verlag GmbH, Hannover

ISBN 3-931 70229-4

Contents

Aqualog Numbers

The numbers preceeding the titles of the photos are identical with the numbers in Aqualog

South American Cichlids II

Introduction

The Dwarf Cichlids of South America are classified in a number of genera and - especially the representatives of *Apistogramma* - in several species-complexes. Not all of them could equally win friends among aquarists. Therefore in this book we concentrate on the representatives of the genera and species of the relating groups *Geophagini* and *Crenicarini*. Particularly the representatives of the genus *Apistogramma*, consisting of almost 60 scientifically described species, belong to the first mentioned group. Due to our own estimations, 50 to 70% of other species have to be added, some of them already being known in the hobby, but their scientific descriptions still have to be done. Thereby, and also because of the number of scarcely known or even unknown species, one can see that the waters of South America are still retaining quite some surprises to the friends of Dwarf Cichlids all over the world. Let us consider only the vast regions around the upper parts of the three huge whitewater rivers Rio Tapajós (with its huge tributaries Rio Juruena and Rio São Manuel [or Rio Teles Pires], Rio Xingú (with the tributaries Rio Iriri and Rio Curuá) as well as the Rio Tocantins (with its huge tributary, the Rio Araguaia). They all come from the central Brazilian massif. The fact that from this, by scientists and hobbyists obviously less frequently visited, area new Dwarf Cichlids were recently discovered, (compare *Apistogramma* spec. ›Peixoto-Azevedo‹ or *Dicrossus* spec. ›Dreipunkt-‹ or ›Tapajós‹), speaks for itself, and the traveller, moving his interest away from Central America to here, could also have further success with discoveries. In a similar way, from the northeast of Brazil until now, only *A. piauiensis* was described scientifically - a huge area where travelling aquarists should take a look. Even from the tropical parts of Perú, which in the recent years brought aquarists a number of new species, more discoveries can be expected (Rio Marañon, Rio Tigre...). As practice showed us, also from certain areas in the south of Brazil, in Paraguay as well as in the northwest of Argentina (border area to Bolivia), discoveries can still be anticipated. In this respect I am thinking of *A. aequipinnis* which in 1938, AHL described with a few sentences, but which stayed in oblivion until today.

Until today, the genus *Apistogrammoides* has remained monotypic (i.e. containing only one species). The same is true for the genus *Taeniacara*.

Since the American authors MYERS & HARRY in 1948, described *Apistogramma ramirezi*, the discussions about the generic classification and distribution of these fish did not come to a rest. The first specimens were imported from the savannah areas of Venezuela to Florida (importer Ramirez). But in the beginning, their exact location was not told by the collector Hermann Blass, and therefore in the scientific publication it could not be given. The reproductive behaviour was unknown either, so that a punctual differentiation based on the fact that this particular species is a substratum – spawner was impossible.

The German scientist Dr Karl-Heinz LÜLING who at that time was working as an ichthyologist in Bolivia, sent a preserved specimen from the low-lying east of this country to the former director of the VDA fish destination department, Mr Hermann MEINKEN. He thought it to be *A. ramirezi* which was erroneously verified by MEINKEN (1967) (»... obviously this is in fact a fully or at least almost fully adult male of *Apistogramma ramirezi* MYERS &

HARRY, 1948, which differs from the original description only in very few measurements. The differences are so faint that thence not even the creation of a subspecies or a well defined variety would be appropriate«).

Towards the end of his explanation, MEINKEN allowed himself to comment, »It can easily be supposed as true that Ramirez and Blass did a 500 miles trip in Venezuela. It is not difficult to believe them. But they did not catch *Apistogramma ramrezi* in Venezuela!«

Protests against this statement were not be missed and already one year later MEINKEN had to correct himself. He did this also by not leaving the protestor untouched, a Mr Fischer from Hamburg: »... if you only would have spoken out earlier...« (1968). The way the transfer into a new genus developed will be shown in the sector "....about families, associated species relationships... ". So far about *Geophagini*.

The relating groups of the *Crenicarini* is poorer in species because out of the three associated genera *Crenicaria*, *Dicrossus* and *Mazarunia,* the former two contain two species each, the latter only one species. Two more *Dicrossus* forms are known, yet their scientific descriptions have to be published. It may be that the old generic name *Dicrossus* is unknown to many aquarists. But it was already created by STEINDACHNER in 1875, whereas in his work dating from 1905, REGAN considered it a synonym of *Crenicara* - an opinion which was accepted by most scientists, until KULLANDER separated the deep bodied *Crenicara* species from the elongated ones of the genus *Dicrossus* by giving back validity to the former generic name in 1990. The old/new generic name was immediately introduced into aquaristic literature, but this does not mean it is noncontroversial.

The mostly unknown representatives of the genus *Mazarunia* come from an isolated area in Guyana (former ›British Guiana‹) in the northeast of South America. These are specimens which Patrick DE RHAM collected and preserved in October 1976.

In order to explain the data of the length and distance in the text, the following table contains examples of conversion from the metric system to GB and US measurements.

Metric System		GB & US
10 millimetres (mm)	= 1 centimetre (cm)	= 0.3937 inches (in)
100 centimetres	= 1 metre (m)	= 39.37 inches or 1.094 yards (yd)
1000 metres	= 1 kilometre (km)	= 0.62137 miles or about 5/8 miles
GB & US		**Metric System**
1 inch (in)		= 25.3995 millimetres (mm)
12 inches	= 1 foot (ft)	= 30.479 centimetres (cm)
3 feet (ft)	= 1 yard (yd)	= 0.9144 metres (m)
1760 yards	= 1 mile	= 1.6093 kilometres (km)

The Habitats of South American Dwarf Cichlids

The distribution of the two mentioned Dwarf Cichlids groups reaches from the North of the South American subcontinent, up to the Argentinian border into the central part of this country. In 1979, in northern Argentina, near the cities Corrientes and Resistencia, by the Rio Paraná, *A. borellii* could be proved (MAYLAND, 1980). For *A. trifasciata*, LINKE & STAECK (1992) indicate the surroundings of the city of Santa Fé by the Rio Paraná as the southernmost location. In Central America, there are no Dwarf Cichlids.

As should be known to most aquarists, the habitats in South America have very soft water which only on the slopes of the Andes is slightly harder. Considering the habitats of Dwarf Cichlids in a more detailed manner, one finds that they can be found mostly in the areas, where the ground is covered by tropical rainforest. Unfortunately, more and more roads cross the country. People are drawn there and subsequently these regions are cleared for farm land and pasture, the cultivation threatening the habitats of these fishes. On the other hand, we know from the northeast of Colombia and the adjacent area of the llanos of Venezuela, which are completely dominated by savannah areas, that under certain circumstances Dwarf Cichlids can also occur and survive in zones like these. Here, as well as in the cases of the reported catches in the north of Argentina, the small Cichlids live in oasislike appearing ponds - or vast lagoons where plants are mostly present, but do not necessarily occur.

Against larger and voracious cohabitants, Dwarf Cichlids can only partially defend themselves, whereas males among each other fight frequently - seen in an aquarium that is too small - and threaten each other with a dropped lower mouth so that the head appears larger, and spread gill covers, if the territorial boundaries are not respected, or are penetrated. Due to their relatively small size, security for these specimens is important. After having moved into a new territory, their first interest is the possession of a small shelter, a small cave or a similar place to hide. There is no lack of this in the periphery of the waters in the tropical rainforest, because sufficient amounts of wood that is twisted together - partially covered with leaves - are on the bottom. If the areas of shallow water have small stony bays these are also tended to be used. These days, with the fairly advanced construction of roads and rough lanes, tubes of corrugated iron with various diameters are used for rainwater drainage. Even in these we found Dwarf Cichlids. Unfortunately, while col-

Biotope: Rio Parurú,
a tributary of the lower Rio Tocantins, downstream the city of Cametá, habitat of the brown Discus.
Date: 25.5.1986
Time: 15.00 o'clock
Water movement: Rapidly flowing. At high tide, the water of the river is damed beyond this river bed.
Water depth: In the Discus biotope there is no shallow shore. One cannot stand.
Water colour: Black water with sediment layers on the bottom of the river bed. Water sample light tea coloured

Laboratory results:

pH-value	5.78
Electrical conductivity (25°C), µS/cm	20
Acid capacity up to pH 8.2 (p-value), mmol/l	n.p.
Acid capacity up to pH 4.3 (m-value), mmol/l	0.12
As carbonate hardness, °d	0.3
Total Alkalis (Ca^{2+}/Mg^{2+}), mmol/l	0.02
As total hardness, °d	0.1
Calcium (Ca^{2+}), mg/l	0.55
Magnesium (Mg^{2+}), mg/l	0.26
Total Iron ($Fe^{2+/3+}$), mg/l	0.37
Manganese (Mn^{2+}), mg/l	n.p.
Ammonium (NH_4^+), mg/l	0.04
Nitrite (NO_2^-), mg/l	0.10
Nitrate (NO_3-), mg/l	2
Chloride (Cl^-), mg/l	4
Sulphate (SO_4^{2-}), mg/l	10
Total Phosphate (PO_4^{3-}), mg/l	0.22
Hydrogen Carbonate (HCO_3^-), mg/l	7
Carbonate (CO_3^{2-}), mg/l	n.p.
Oxidation (Potassium Permanganate $KMnO_4$ consume, mg/l	14

n.p. = not provable

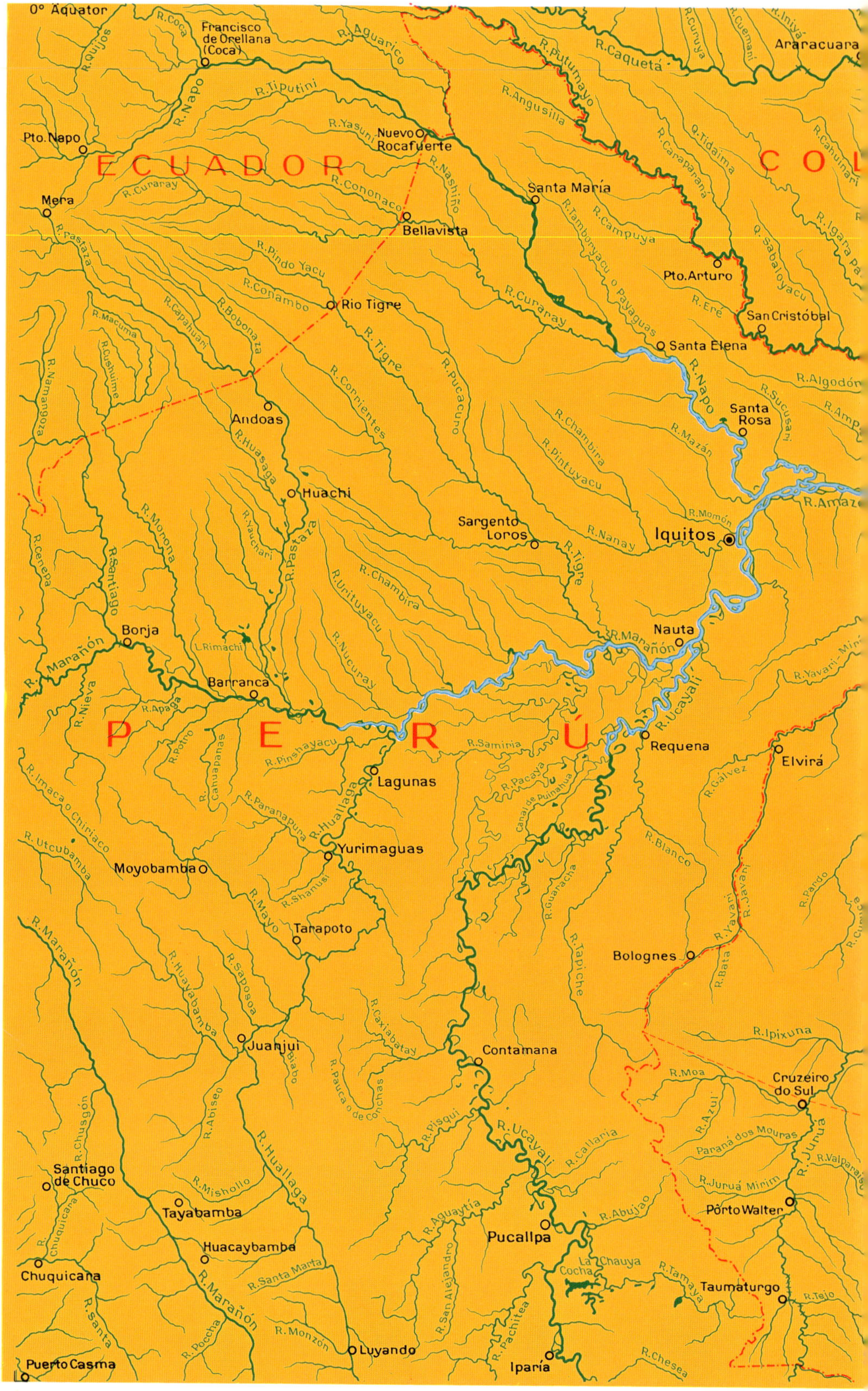

0° Äquator
ECUADOR
CO
PERÚ
Francisco de Orellana (Coca)
Pto. Napo
Mera
Nuevo Rocafuerte
Bellavista
Rio Tigre
Andoas
Huachi
Santa María
Pto. Arturo
San Cristóbal
Santa Elena
Santa Rosa
Araracuara
Iquitos
Sargento Loros
Nauta
Borja
Barranca
Requena
Elvirá
Lagunas
Yurimaguas
Moyobamba
Tarapoto
Juanjui
Bolognes
Contamana
Cruzeiro do Sul
Pôrto Walter
Pucallpa
Taumaturgo
Santiago de Chuco
Tayabamba
Huacaybamba
Chuquicana
Puerto Casma
Luyando
Iparía
La Chauya
Cocha
R. Coca
R. Quijos
R. Napo
R. Aguarico
R. Tiputini
R. Yasuni
R. Curaray
R. Cononaco
R. Nashiño
R. Putumayo
R. Caquetá
R. Cunuya
R. Cuemani
R. Inivá
R. Angusilla
R. Carapanana
Q. Tidaima
R. Cahuinari
R. Igara Paraná
R. Tamboryacu o Payaguas
R. Campuya
Q. Sabaloyacu
R. Eré
R. Algodón
R. Ampiyacu
R. Sucusari
R. Mazán
R. Pastaza
R. Pindo Yacu
R. Conambo
R. Capahuari
R. Bobonaza
R. Macuma
R. Cushuime
R. Namangoza
R. Tigre
R. Corrientes
R. Pucacuro
R. Chambira
R. Pintuyacu
R. Nanay
R. Momón
R. Amazonas
R. Huasaga
R. Morona
R. Yaucharí
R. Cenepa
R. Santiago
R. Urituyacu
R. Nucuray
L. Rimachi
R. Marañón
R. Nieva
R. Apaga
R. Potro
R. Cahuapanas
R. Pinshayacu
R. Samiria
R. Pacaya
Canal de Puinahua
R. Ucayali
R. Yavari-Mirim
R. Gálvez
R. Blanco
R. Imaca o Chiriaco
R. Utcubamba
R. Paranapura
R. Huallaga
R. Shanusi
R. Mayo
R. Guaracha
R. Tapiche
R. Yavari
R. Javari
R. Pardo
R. Curuça
R. Batá
R. Huayabamba
R. Saposoa
R. Caxiabatay
R. Biabo
R. Pauca o de Conchas
R. Ipixuna
R. Moa
R. Azul
Paraná dos Mouras
R. Juruá
R. Juruá Mirim
R. Valparaíso
R. Abiseo
R. Pisqui
R. Callaria
R. Chusgón
R. Mishollo
R. Aguaytía
R. Abujao
R. Tamaya
R. Tejo
R. Chuquicara
R. Santa
R. Santa Marta
R. San Alejandro
R. Pachitea
R. Poccha
R. Monzón
R. Chesea

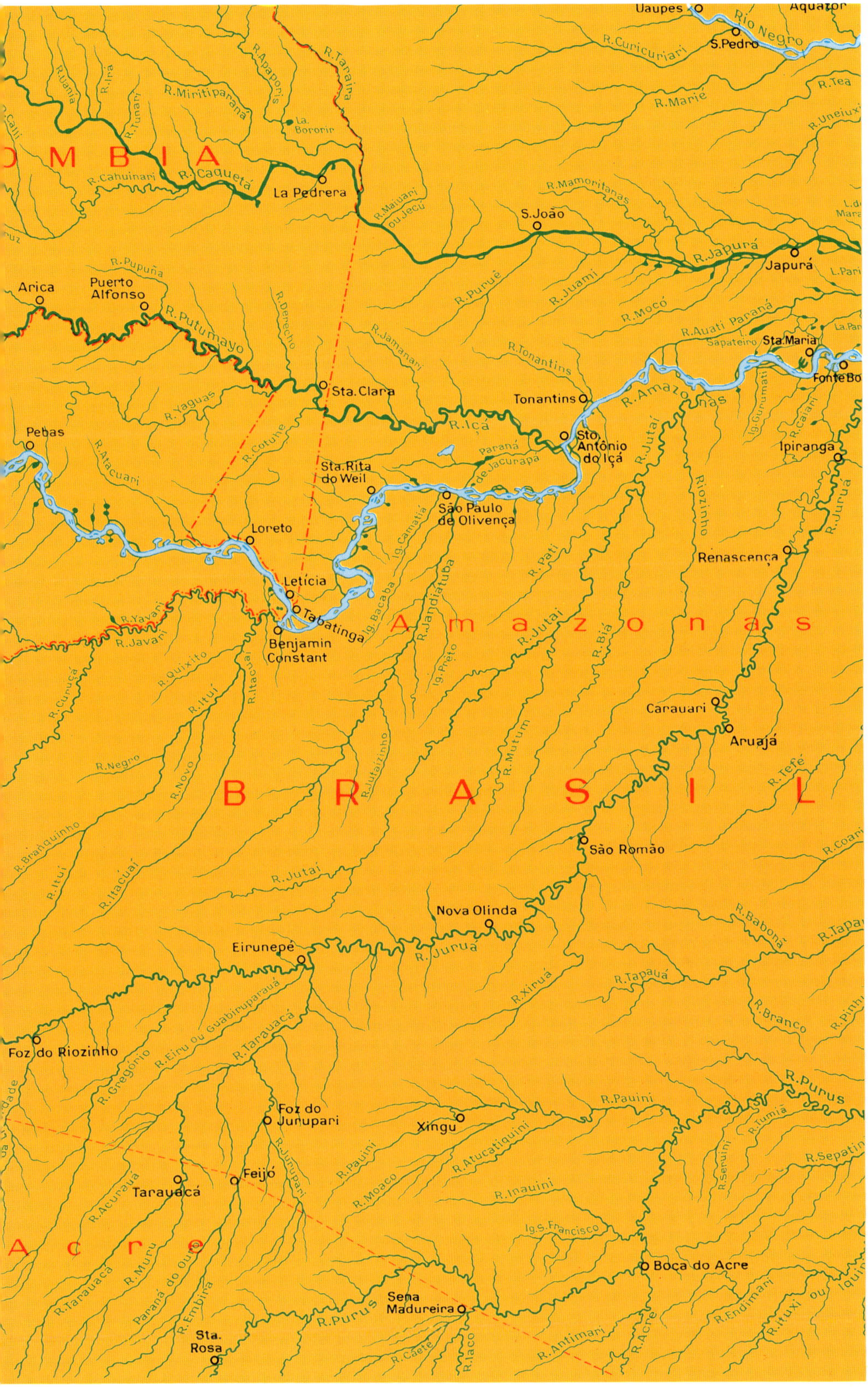

Uaupes
Rio Negro
S.Pedro
R.Curicuriari
R.Tea
R.Marié
R.Uneiuxi
R.Uania
R.Ira
R.Tunani
R.Miritiparana
R.Apaporis
R.Tanaina
La. Bororin
OMBIA
R.Cahuinari
R.Caquetá
La Pedrera
R.Maiuari ou Jecú
R.Mamoritanas
S.João
R.Japurá
Japurá
L.Pari
R.Pupuña
Arica
Puerto Alfonso
R.Putumayo
R.Derecho
R.Puruê
R.Juami
R.Mocó
R.Auati Paraná
Sapateiro
Sta.Maria
Fonte Bo
R.Jamanari
R.Tonantins
Sta.Clara
Tonantins
R.Amazonas
R.Yaguas
R.Içá
Pebas
R.Atacuari
R.Cotune
Sta.Rita do Weil
Paraná de Jacurapá
Sto. Antônio do Içá
R.Jutaí
Riozinho
Ig.Curumati
R.Caiani
Ipiranga
São Paulo de Olivença
R.Juruá
Loreto
Ig.Camatiã
R.Pati
Renascença
Letícia
Tabatinga
Ig.Bacaba
R.Jandiatuba
A m a z o n a s
R.Yavari
R.Javari
Benjamin Constant
R.Itacuaí
R.Quixito
R.Jutaí
R.Biá
Ig.Preto
R.Curuçá
R.Ituí
R.Mutum
Carauari
Aruajá
R.Negro
R.Novo
R.Jutaizinho
R.Tefé
B R A S I L
R.Branquinho
R.Ituí
R.Itacuaí
R.Jutaí
São Romão
R.Coari
Nova Olinda
R.Baboná
R.Tapauá
Eirunepé
R.Juruá
R.Xiruã
R.Tapauá
R.Eiru ou Guabiruparauá
R.Tarauacá
R.Branco
Foz do Riozinho
R.Gregório
R.Purus
R.Pauini
Foz do Junupari
Xingu
R.Tumiã
R.Seruini
R.Sepatini
R.Pauini
R.Atucatiquini
R.Junupari
Feijó
Tarauacá
R.Acuraua
R.Moaco
R.Inauini
Ig.S.Francisco
A c r e
R.Muru
R.Tarauacá
Paraná do Ouro
R.Embira
Boca do Acre
Sena Madureira
R.Purus
R.Endimari
R.Ituxi ou Iquiri
Sta. Rosa
R.Caeté
R.Iaco
R.Antimari
R.Acre

Biotope: Rio Zaraíba,
near the city of Benjamin Constant/west Brazil, but the river is actually in Perú
Date: 6.3.1987
Time: 10.30 o'clock
Water movement: Slowly flowing
Water depth: Not established
Water colour: Light tea colour

Laboratory results:

pH-value	6.74
Electrical conductivity (25°C), µS/cm	85
Acid capacity up to pH 8.2 (p-value), mmol/l	n.p.
Acid capacity up to pH 4.3 (m-value), mmol/l	0.73
As carbonate hardness, °d	2.05
Total Alkalis (Ca^{2+}/Mg^{2+}), mmo/l	0.39
As total hardness, °d	2.19
Sodium (Na^+), mg/l	2.9
Potassium (K^+), mg/l	1.3
Calcium (Ca^{2+}), mg/l	13
Magnesium (Mg^{2+}), mg/l	1.6

Total Iron ($Fe^{2+/3+}$), mg/l	0.30
Manganese (Mn^{2+}), mg/l	n.p.
Ammonium (NH_4^+), mg/l	0.03
Nitrite (NO_2^-), mg/l	n.p.
Nitrate (NO_3^-), mg/l	n.p.
Chloride (Cl^-), mg/l	3
Sulphate (SO_4^{2-}), mg/l	6
Total Phosphate (PO_4^{3-}), mg/l	n.p.
Hydrogen Carbonate (HCO_3^-), mg/l	45
Oxidation a) As Potassium Permanganate $KMnO_4$ consume, mg/l	6
b) As oxygen (O_2), mg/l	1.5

n.p. = not provable

lecting such fish one finds their enemies time and time again. They not only attack the adult specimens, but above all, they eat also their fry. Among the main predators are the characid fish of the genus *Hoplias* (Family *Erythrinus*) and the Cichlids of the genus *Crenicichla*. Besides these, the Piranhas should not be neglected as well as nocturnal predatory catfish and knife fish. Specimens of such species are in the collector's net more often than he likes it. If in the obviously limited habitat the mentioned hiding places are missing, the Cichlids occasionally hide in the overhanging vegetation of the banks. In smaller ponds, the so-called rest water pools, the selected habitat can turn out to be a deadly trap because in some of these ponds the water evaporates completely during the dry season.

It has already been mentioned that in South America the water on average is relatively soft. To most of the Dwarf Cichlid keepers it should be also known that basically in the habitats of this subcontinent three different types of water are found (white water, black water and clear water) and the corresponding converging zones. Many of the *Apistogramma* species that are described as the most beautiful ones, come from regions in Central America or from those in the upper Rio Orinoco where the amount of black water is dominant. These fish generally place higher demands on their keeper. On the other hand, the demands we know from the Dwarf Cichlids of the marginal areas of the Amazon basin (*A. cacatuoides*, *A. caetei*, *A. eunotus*, *A. linkei*, *A. steindachneri*, *A. uruteagai*) or from the drainage of the Rio Paraguay and the Rio Paraná (*A. aequipinnis*, *A. borelli*, *A. commbrae*, *A. trifasciata*, *Microgeophagus altispinosus*) are lower.

In judging the water quality of most of the South American rivers (in as far as they do not directly pass a major city) one must not consider only the softness of the water. The extreme low concentration of nitrate is not only important from the view of a Discus breeder, but should also be taken into consideration for the reproduction of Dwarf Cichlids. Moreover, one should also give up the common opinion that only black water contains many acidic particles.

According to specifications in literature, the annual average temperature at least for the habitats in central Amazonia, is 26°C. In many breeding reports, one finds that for stimulation of reproduction the water temperature has to be raised several degrees. At least for reproducing some species (for example *A. nijsseni*, see page 93), breeders could find out that raising the temperature or also lowering the average water temperature in the breeding tank, influence the sex of the offspring: Raising the temperature increases the production of male specimens, lowering the temperature favours females. But astonishingly, mass screening (Römer, 1996, pers. comm.) showed that the results in Dwarf Cichlids (*A. caetei*, *A. piauiensis*) whose occurrence

Apistogramma group above foliage in the natural location in the Rio Xingú drainage.

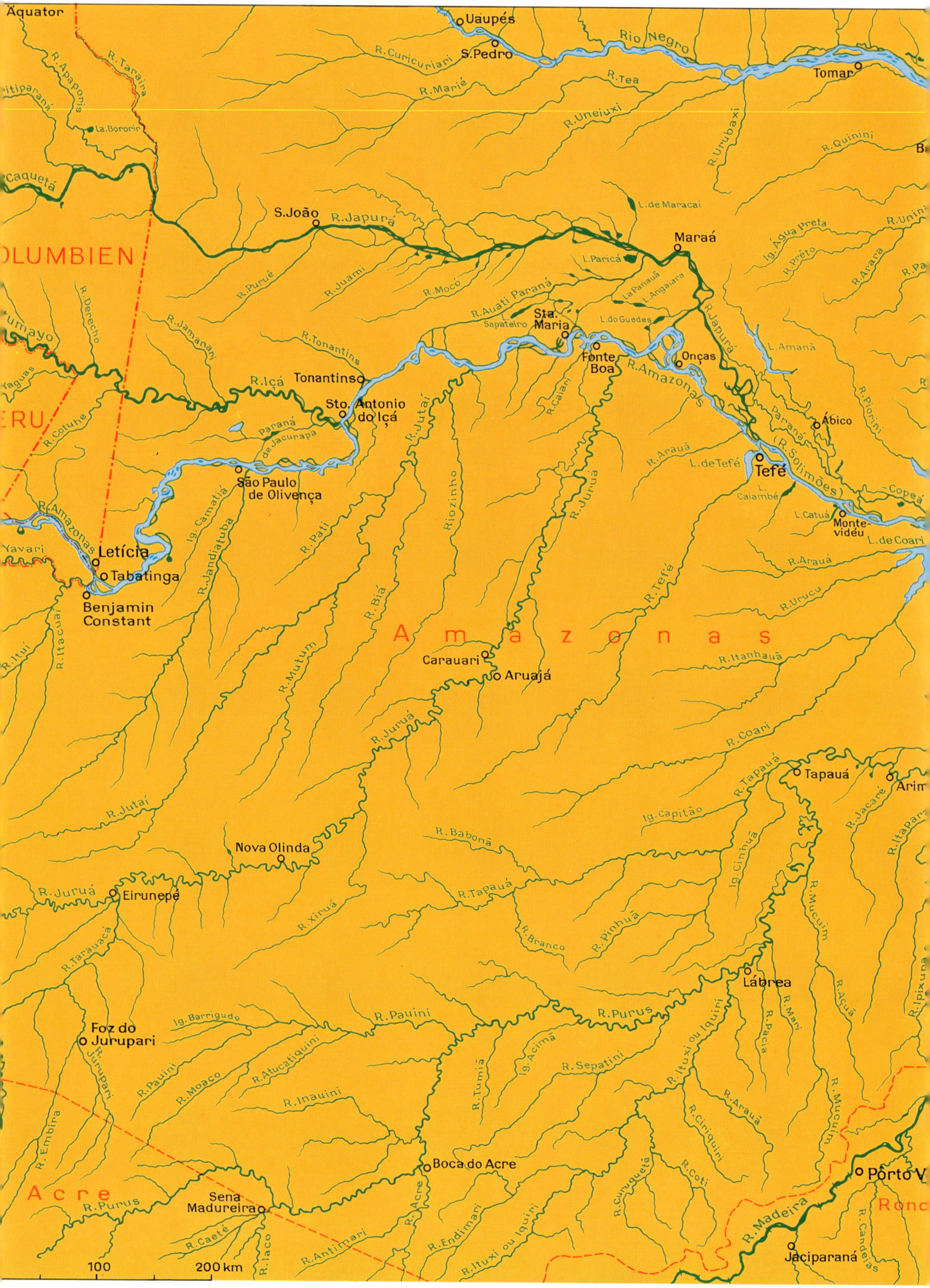
Äquator
Uaupés
S.Pedro
Rio Negro
R.Curicuriari
R.Marié
R.Tea
Tomar
R.Uneiuxi
R.Urubaxi
R.Quinini
R.Tarairá
R.Apaporis
La.Bororin
Caquetá
S.João
R.Japurá
L.de Maracaí
Maraá
KOLUMBIEN
R.Puruê
R.Juami
R.Mocó
L.Paricá
Ig.Água Preta
R.Prêto
R.Arara
R.Derecho
R.Jamanari
R.Auati Paraná
Sapateiro
Sta. Maria
L.do Guedes
L.Amanã
R.Tonantins
Tonantins
R.Içá
Sto. Antonio do Içá
Fonte Boa
Onças
R.Amazonas
R.Japurá
R.Cauiari
R.Jutaí
Paraná de Jacurapá
Ábico
R.Cotuhe
Yaguas
PERU
São Paulo de Olivença
R.Juruá
R.Araua
L.de Tefé
Tefé
Paraná (R.Solimões)
L.Caiambé
L.Catuá
Monte-vidéu
L.de Coari
Copeá
R.Amazonas
Riozinho
Letícia
Tabatinga
Benjamin Constant
Yavari
Ig.Camatiá
R.Jandiatuba
R.Pati
R.Biá
R.Tefé
R.Araua
R.Urucu
Amazonas
Carauari
Aruajá
R.Itanhauã
R.Ituí
R.Itacuaí
R.Mutum
R.Juruá
R.Coari
R.Tapauá
Tapauá
R.Jacaré
R.Itaparaná
R.Jutaí
Ig.Capitão
R.Babonã
Nova Olinda
Ig.Cinhuã
R.Juruá
Eirunepé
R.Tapauá
R.Xiruá
R.Mucuim
R.Branco
R.Pinhuã
R.Tarauacá
Lábrea
R.Mari
R.Pacia
R.Açuã
R.Ipixuna
Ig.Barrigudo
R.Pauini
R.Purus
Foz do Jurupari
R.Jurupari
Ig.Acimã
R.Sepatini
R.Ituxi ou Iquiri
R.Pauini
R.Moaco
R.Atucatiquini
R.Inauini
R.Tumiã
R.Ciriquiri
R.Arauã
R.Mucuim
R.Embira
Boca do Acre
R.Curuquetê
R.Coti
Acre
R.Purus
Sena Madureira
R.Acre
Pôrto V
Ronc
R.Madeira
R.Caeté
R.Iaco
R.Antimari
R.Endimari
R.Ituxi ou Iquiri
Jaciparanã
R.Candeias
100
200 km

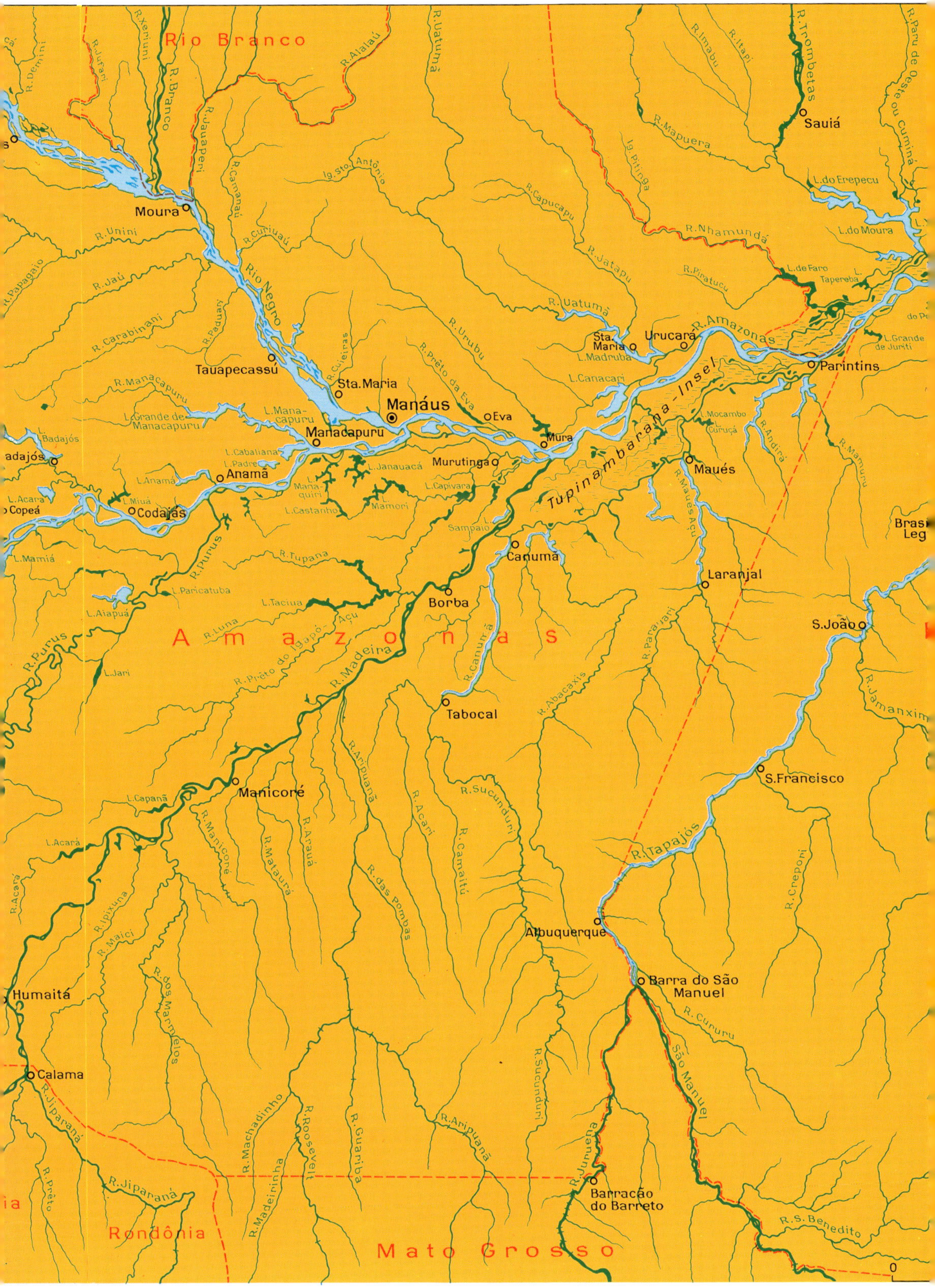

Rio Branco
R.Demini
R.Jufari
R.Xeriuini
R.Branco
R.Jauaperi
R.Alalaú
R.Uatumã
R.Imabu
R.Itapi
R.Trombetas
Sauiá
R.Paru de Oeste ou Cuminá
R.Mapuera
R.Camanaú
Ig.Sto.Antônio
Ig.Pitinga
L.do Erepecu
L.do Moura
Moura
R.Unini
R.Curiuaú
R.Capucapu
R.Nhamundá
R.Jatapu
R.Pinatucu
L.de Faro
L.Tapereba
R.Papagaio
R.Jaú
Rio Negro
R.Uatumã
R.Carabinani
R.Padauri
R.Cuieiras
R.Urubu
R.Prêto da Eva
Sta.Maria
L.Madruba
Urucará
R.Amazonas
L.Grande de Juriti
Parintins
Tauapecassú
Sta.Maria
R.Manacapuru
Manáus
L.Canacari
L.Grande de Manacapuru
L.Manacapuru
Eva
Tupinambarana-Insel
L.Mocambo
Curuçá
Badajós
Manacapuru
Mura
R.Andirá
R.Mamuru
L.Cabaliana
L.Padre
L.Janauacá
Murutinga
Maués
Anamã
L.Anamã
L.Manaquiri
L.Capivara
R.Maués Açu
L.Acara
L.Miuá
Copeá
Codajás
L.Castanho
Mamori
Sampaio
R.Purus
L.Mamiá
R.Tupana
Canuma
Laranjal
L.Paricatuba
L.Aiapuá
L.Tacua
Borba
S.João
R.Luna
R.Prêto do Igapó-Açu
Amazonas
R.Canumã
R.Paraurari
R.Purus
L.Jari
R.Madeira
R.Abacaxis
R.Jamanxim
Tabocal
R.Aripuanã
R.Sucunduri
S.Francisco
L.Capanã
Manicoré
R.Acari
R.Manicoré
R.Mataurá
R.Arauá
R.Camaitú
R.Tapajós
L.Acará
R.Acará
R.Ipixuna
R.das Pombas
R.Crepori
R.Maici
Albuquerque
Humaitá
R.dos Marmelos
Barra do São Manuel
R.Cururu
São Manuel
R.Sucunduri
Calama
R.Jiparaná
R.Machadinho
R.Roosevelt
R.Guariba
R.Aripuanã
R.Juruena
R.Prêto
R.Jiparaná
R.Madeirinha
Barracão do Barreto
R.S.Benedito
Rondônia
Mato Grosso

Biotope: Lago Berurí,		**Laboratory results:**			
besides the River Purús		Electrical conductivity (25°C), µS/cm	63	Total Iron ($Fe^{2+/3+}$), mg/l	0.17
Date: 3.10.1986		Acid capacity up to pH 8.2		Manganese (Mn^{2+}), mg/l	<0.02
Time: 15.00 o'clock		(p-value), mmol/l	<0.05	Ammonium (NH_4^+), mg/l	0.05
Water movement: Sluggishly flowing		Acid capacity up to pH 4.3		Nitrite (NO_2^-), mg/l	<0.02
Water depth: Not established		(m-value), mmol/l	0.54	Nitrate (NO_3^-), mg/l	<1
Water colour: Light tea brown		As carbonate hardness, °d	1.5	Chloride (Cl^-), mg/l	3
		Total Alkalis (Ca^{2+}/Mg^{2+}), mmol/l	0.22	Sulphate (SO_4^{2-}), mg/l	9
Field measurements:		As total hardness, °d	1.2	Total Phosphate (PO_4^{3-}), mg/l	0.09
pH-value	7.6720	Calcium (Ca^{2+}), mg/l	6.8	Hydrogen Carbonate (HCO_3^-), mg/l	33
Water temperature	31.3°C	Magnesium (Mg^{2+}), mg/l	0.20	Oxidation (Potassium Permanganate	
Air temperature	32.0°C			$KMnO_4$ consume), mg/l	324

in nature could be called ›close to the atlantic shore‹ brought opposite figures. Reversed conditions showed: Raising the temperature = more female offspring, lowering the temperature = more male offspring. By the way, similar consequences as those mentioned first are also known to breeders of crocodile, respectively caiman species, and is also known to workers in the breeding stations of marine turtles. As written previously, corresponding findings for Dwarf Cichlids are added to the breeding information of specific species that follow later in this work.

The water tests published in this book - professionally judged results from waters close to the habitats of the corresponding species, partially combined with our own field measurements - can present only a limited look at the water chemistry in the moment the sample was taken. These shall and can also give valuable hints to breeders. Today, there are still many question marks in the habitats of ›our‹ Dwarf Cichlids, for which concrete answers are difficult to find. With purely theoretical thinking at meetings of clubs and associations not all of these questions will be answered.

The following table contains examples of conversion from °C (degrees Celsius) °F (degrees Fahrenheit) and vice versa.

18°C = 64.4°F	64°F = 18.8°C
19°C = 66.2°F	66°F = 19.9°C
20°C = 68.0°F	68°F = 20.0°C
21°C = 69.8°F	70°F = 21.1°C
22°C = 71.6°F	72°F = 22.2°C
23°C = 73.4°F	74°F = 23.3°C
24°C = 75.2°F	76°F = 24.4°C
25°C = 77.0°F	78°F = 25.5°C
26°C = 78.8°F	80°F = 26.6°C
27°C = 80.6°F	82°F = 27.7°C
28°C = 82.4°F	84°F = 28.8°C

South American Dwarf Cichlids in the Aquarium

When one has the chance to take a look into the aquaria of many aquarists - for example during a homeshow of a club - one will soon learn that Dwarf Cichlids are only kept in special aquaria with an average length of 80 cm (100 cm are better) and a width of 40 to 50 cm, by those aquarists whose particular intention is reproduction. But if the aquaria are large enough (100 cm length and more) and richly decorated with plants and hiding places made of wood and/or stone, the Dwarf Cichlids - especially the easier species - can also be kept in a community tank. One requirement for this is that the needs of the small Cichlids are taken into consideration. Regular constant partial water change, careful handling of chemicals [if ever!] and absence of predatory cohabitants. Without these requirements, many specimens remain shy or are irritated by shocks which means that they flee as soon as somebody is approaching the tank. Of course, larger predators consider the small Cichlids as prey also in the aquarium. Very suitable is a socialization with small Tetras or lifebearers, catfish associated with the *Loricaria* species and representatives of the smaller *Corydoras* species.

Depending on what the aquarium is used for, its water has two functions. It has to fulfil the needs of the fish and - particularly in community tanks - those of the aquatic plants. Most of the Dwarf Cichlids come from biotopes with water of a more or less acidic pH (6.8 to 6.0). Only few rivers, mostly near the Andes, have water with a pH only a few points above the neutral level (7.0). As in acidic water, and water with insufficient minerals, the plants lack important nutriants, they do not grow particularly well so that an aquarist keeping Dwarf Cichlids in community tanks with aquatic plants has to find alternatives. For breeders it has to be said that not only a change of the water temperature, as it was mentioned in the previous chapter, can be the reason for too many male or female offspring, but in this concern, the pH also plays an important role: a drop in the pH value = more males, a rise in the pH value = more females. Results from researching the causes were published for representatives of the genus *Apistogramma* by RÖMER & BEISENHERZ (1995), where the correlation between temperature, respectively pH value and the resulting percentage of male and female offspring was investigated.

›The ideal tropic water‹ does not flow out of every tap. Often it has to be treated. But the softer the water is, the worse its pH can be affected. If somebody wants to treat his tap water with ion exchangers or even by reverse osmosis, it is the simplest way to get into contact with a friendly Discus breeder. Especially concerning reverse osmosis, in his book ›Praxis Diskusaquarium‹ (same publisher) MAYLAND has discussed this subject. Contrary to ion exchange, reverse osmosis also takes traces of the fungicides, herbicides and pesticides out of the water, these being harmful to the reproduction of sensitive fish. In both cases, the tap water which was lowered in hardness and acidity in the ways mentioned before, cannot be used without being treated a second time. In case of a perfectly working water treating system, certain chemicals were completely taken out so that in the actual state, it has to be called ›dangerous to life‹. Only by adding certain mineral combinations, it regains its suitability as good aquarium water. If the water which was softened and brought to a neutral pH value is mixed with tap water, it should be considered that once more chemicals that inhibit reproduction - even in lower concen-trations - are brought into the aquarium water.

In order to make water of a more or less neutral pH value more acidic, a filtration through peat is recommended. Corresponding peat additives in various concentrations are offered in aquarium shops.

Besides hiding places Dwarf Cichlids also love an environment that is not too bright. Hence, for the setup of an aquarium it is important that besides a dark bottom also dark covers of the back and lateral screens are chosen. Also for this purpose, aquarium shops offer material. As these slabs (made of polyurethane foam) have enclosed air bubbles, they show a proper buoyancy. Therefore it is the best to glue them with a few drops of silicon. The manufacturer's information has to be followed; avoid access to unauthorized people. With some skill, even plants can be cultivated on these slabs.

Monogamy means a relation to one partner only. But for how long? For the entire life or only for one reproductive period? Contrary to this, polygamy stands for a synchronic relation to several partners. Interested hobbyists know this best from mouthbreeding Cichlids of the East African lakes. In these fish, between the two partners there is no life long mating. The more colourful males constantly show themselves as sexually active and mate with every female specimen of the same species. Some females are organized in groups and present themselves to the males only when they are ready for mating which means that the moment of spawning has come. On the other hand, the strong intraspecific aggression forces the males to defend their territory against rivals of the same species, once it has been occupied.

With the representatives of Dwarf Cichlids dealed in this publication, we find similar conformity. The more colourful attractive male specimens control their territory. Only when the aquarium is quite big and has a rich choice of cavelike hiding places, besides a ›harem‹ of females, several males can be put in (comp. *A. diplotaenia*). In biology (ethology = behaviour of animals), the term ›harem‹ describes a group of females that is looked after and bound together by one male which additionally prevents other males from reproducing with ›its‹ females (HEYMER, 1977; TEMBROCK, 1978). Indeed, the partnership does not last for life, in some cases it only lasts up to the fertilization of the latest eggs which can also be caused by the fact that a female looking after a brood chases its partner from the cramped breeding area.

For building various breeding caves, stones, flags, wood (cork) or even synthetic material can be used. As females of some Apistogramma species can be quite small, they choose just as small, bottom-oriented spawning sites. They will readily accept, for example, a plastic tube buried diagonally into the substrate, so that only the entry remains visible. Finding a such prepared, ideal spawning site, the female can also give way to her tendency to make the entry even narrower. For the satisfaction of his own curiosity, the aquarist can arrange the cave or the tube in a way so that through the front screen of the aquarium - if necessary with the help of an electric torch - he can look into the cave and can satisfy himself of the progression of a possible brood.

The coming Dwarf Cichlid keeper cannot start thinking about feeding along with the corresponding provision of food soon enough. It may be that in a community tank specimens of one or another species turn out to be happy with the common flake food - however this food is not sufficient permanently because in the natural habitat, fleshy food consisting of insects, worms and other small shrimps, is part of their main nutrition. Their digesting organs are adapted to that. Unfortunately, for us aquarists it is not possible to get ›original South American food‹, but insects, worms and small shrimps can also be found in other waters.

First of all a few words about the problem food ›tubifex‹: these worms belonging to the *Oligochaetae* live in mud. Their habitat in Southeast Asia (mostly from Taiwan) is chemically very much polluted, - even if they are imported in a deep frozen or freeze dried state - which results in poison in the worms. They even tolerate waters where phenoles and heavy metals are in solution and therefore their body tissues accumulate these chemicals without taking harm themselves. Even rinsing of any kind does not help!

Semi-adult Dwarf Cichlids - as well as fry - also eat brine shrimp nauplii. If you have the possibility to catch *Daphnia*, *Cyclops* and the various kinds of insect larvae in uncontaminated waters - whereby the contemporary regulations have to be taken into account (the department of conservation will inform you) - you can be happy to have such a source of life food. Among the *Crustacea*, routined breeders obviously prefer *Cyclops* to *Daphnia*. Moreover *Cyclops* also mean an excellent spawning stimulation for females with eggs. If these food resources ever fail, in case of emergency most Dwarf Cichlids can also be fed with deep frozen food. Certain specimens - among these, particularly males - tend to a relatively gigantic growth, but first of all to obesity. Also if you love to observe your fish while eating, you should consider that too big and above all, too deep bodied speci-mens do not correspond any more to what we know from the natural habitats in South America. Besides, in most cases they can become useless for reproduction.

Feeding the fry is relatively easy. Already directly after they have started to swim around, they are able to take the larvae (= nauplii) of the brine shrimp (*Artemia salina*). But during the first days of their lives, it is recommended bringing food closer to the fry with a glass pipette (or a thin glass tube) so that it is not too widely dispersed in the aquarium. The shrimp larvae should be sieved before being given to the fry, at least they must be freed from the saltwater and rinsed with freshwater.

The diseases Dwarf Cichlids contract can be the same known from other aquarium fish, for example: dropsy, bacterial fin infections (mainly in younger specimens), *Columnaris* disease (bacterial infection of the skin, decomposition of fin edges so that the fin rays are exempted) and of course also *Ichthyophthirius* as well as other common diseases known within the hobby.

Most, if not all, of the Dwarf Cichlids are kept in special aquaria and therefore do not have too much contact with many fish, the risk of infection as known from community tanks is less frequently given. Therefore when buying the fish it is particularly recommended to watch for a healthy and well fed impression. Even if there is only the slightest suspect of illness, one shall use the well-tried method and first of all supervise the novices in a quarantine tank. As the fish can not tell us about his indisposition, the hobbyist has to make a rough diagnosis which however should be based on a good technical book (cp. UNTERGASSER).

...about Families, Groups of Associated Species, Genera, Species, Subspecies, Races, Varieties and Populations

It should be known to most Dwarf Cichlid keepers that the fish presented here belong to the family *Cichlidae*. The number of species classified in this family (between 1400 and 1500) is vast, and Cichlids do not exist only in South America, but also in Central America, in southern North America, as well as in Africa, on the island of Madagascar and even in southern and western parts of Asia (cp. MAYLAND: ›Cichlids‹, same publisher). From South America, we know about 300 species - but these are only 6 to 10% of the freshwater fauna of South America.

As already mentioned, the Dwarf Cichlids of South America that belong to about one dozen of genera were partitioned into several groups of associated species. These can be divided into four groups, of which the members of the former two groups belong to the most preferred in the hobby:

1. The *Geophaginii*: among these are the representatives of the perhaps best known genus *Apistogramma*. *Apistogrammoides*, *Microgeophagus* (formerly *Papiliochromis*) and *Taeniacara* have to be added.

2. The *Crenicarinii*: this group consists of the members of the genera *Crenicara*, *Dicrossus*, and *Mazarunia*.

3. The *Cichlasominii*: the members of the Dwarf Cichlid genera *Laetacara*, *Bujurquina* and *Nannacara* are added to these relatives of *Cichlasoma*, of which in the past, the former two were mostly classified as members of the genus *Aequidens*.

4. The *Crenicichlinii*: elongated, living mostly at low level, members of the genera *Crenicichla* (here only the ›dwarfs‹ are considered) and *Teleocichla*.

As KULLANDER stated on the occasion of his revised genetic diagnosis in 1989, until now two species of another genus (*Biotoecus*) could not yet be classified. Therefore, in this book we placed these cave spawning species behind *Apistogramma* and *Apistogrammoides*.

If one penetrates deeper into the lower categories of systematics, through the genera, one reaches the species. Apart from the traits that have to be defined: How can one detect the affiliation to a species? About this, scientific literature (MAYR, 1975) informs us as follows: »species are groups of crossbreeding natural populations that with regard to their reproduction, are isolated from other groups«. In this concern, one does not speak about a biological species concept because it relates to biological taxa, but because it is pointed out biologically.

About this, the quoted autor remarks: »Besides, the biological species concept solves the apparent paradox caused by the conflict between the invariability of species (of the natural scientists) and the variability of species (of the evolutionists). It was this conflict that induced LINNAEUS to deny evolution and made DARWIN contest the reality of species.«

Dealing more intensely with Dwarf Cichlids in this concern, one discovers that in the past there were some authors (HASEMAN, FOWLER, MEINKEN) who described subspecies of *Apistogramma* species. With KULLANDER's revision (1980), these subspecies were either considered invalid (*Heterogramma trifasciatum maciliense* HASEMAN, *Apistogramma trifasciatum haraldschultzi* MEINKEN) or they received, in a roundabout way, (KULLANDER & NIJSSEN, 1989, page 82) the status of a valid species (*Apistogramma rupununi*). Scientifically funded works on the status of *A.*

maciliensis as a valid species (comp. Page 145f) are still missing. Are subspecies now ›out of fashion‹?

Already LINNAEUS acknowledged the "varie-tas", the variety, as the only subunit of a species. This was his term for every aberration of a species' type. But indeed for the early taxonomists this caused frequent confusion of individual varieties as well as various kinds of ›races‹. The result: For a period of time the term ›variety‹ was discredited. On the other hand, during the 19th century, the term ›subspecies‹ was used more and more. It should replace the term ›variety‹. Referring to the present use, MAYR indeed writes: »There is no criterion for the definition of the category subspecies that were not artificial. The subspecies is no evolutionary unit, either - except it represents a geographic or genetic isolate at the same time. A subspecies is a comprehension of phenotypically* similar populations of a species which inhabit a geographic part of the distribution area of the species and which differ taxonomically from other populations of the species.

A subspecies can consist of many local populations which - although quite similar to each other - in their completeness differ slightly, genetically and phenotypically; therefore a subspecies is no collective unit. Every local population can be distinguished from any other local population; the presence of these differences can be proved by sufficiently thorough measurements and statistic procedures. It would be paradoxical and would lead to a nomenclatoric chaos if one would want to furnish formally every such a population with a trinom, as this is common for a subspecies. That is why subspecies can only be named if they taxonomically differ from each other, i.e. by sufficient diagnostic morphological traits.«

If a form is named ›race‹, but is otherwise not considered formally as a subspecies, this nonconsideration is also valid in the taxonomic hierarchy.

Populations are variable. For example, we know them from the Cichlids of Lake Tanganjika (for example *Tropheus moorii* and others) as well as those of Lake Malawi (for example *Pseudotropeus elongatus*, *P. zebra* and others). With the Dwarf Cichlids of South America, above all we know the various populations of *Apistogramma agassizii* along the Amazon lowland from the lower Rio Ucayali to the delta (cp. photos).

Corresponding to the variability of populations, MAYR writes, »the description, measurement, and recording of the variation has become one of the main tasks of the taxonomist analysing the taxa on the level of the lower categories. Due to the introduction of the polytypic* species, working on the base of populations did not only lead to a simplification of taxonomy, besides this, it has brought a new approach for the examination to questions of evolution.« In this concern it shall be remarked that all over the world evolution is moving on and on which could result in the fact that in foreseeable time (?) new species develop out of populations that today are still isolated.

Geographic isolations among the Dwarf Cichlids are not only found in the mentioned species *A. agassizii*, but also in many other species, even if remarkable aberrations in colouration and colour pattern occur less frequently. Some of these variants are additionally pictured in this book.

---

* phenotypically means as much as ›referring to the appearance of an organism‹.

* poly = much, more.

Species and Genera

The genus *Apistogramma* REGAN, 1913

A Historic Review

In dependence of KULLANDER's work (1980) which is surely not known to all friends of Dwarf Cichlids, the following chapters shall report about the changeable history of the genus *Apistogramma*. It already started in the 19th Century, when in 1852, the British natural scientist Henry Walter BATES (1825-1892) discovered the first specimen of a Dwarf Cichlid in the Brazilian Rio Cupaí (**not** Rio Cupari, cp. species presentation *Apistogramma taeniata*), which today is still considered the type of the genus *Apistogramma*. BATES sent this only prepared specimen of 42.7 cm length to Albert C.L.G. GÜNTHER (1830-1914), the longtime worker and Director of the Zoological Department of the British Museum of Natural History in London. But GÜNTHER allowed himself ten years time for the scientific description and published it first in volume 4 of his ›Catalogue of the fishes in the British Museum‹ (1862, page 312) as ›*Mesops taeniatus*‹. For this new species as well as for *Geophagus cupido* HECKEL, 1840 GÜNTHER also created a new genus: *Mesops*, for which he found out that it would be very close to *Geophagus*. At this occasion, a generic type for *Mesops* was not defined.

In the years 1865/1866, in the Amazon area, the THAYER expedition took place, which is also known among aquarists, in which among others, also Louis R. AGASSIZ participated. In one of his works (AGASSIZ & AGASSIZ, Reprint 1969, page 184) there is a report about a bigger number of Chromids, mostly Cichlids. More than 600 species - probably mostly young fish - were preserved. They all would have to be classified as members of the present genus *Apistogramma* (which at that time did not yet exist). A part of this material was later ex-amined by the Austrian ichthyologist Franz STEINDACHNER, who consequently in 1875, also realized the scientific description of *A. agassizii* (from Rio Curupirá, Rio Cuda-jas, Rio Puty, Lago Maximo and Lago Manacapurú). All these areas are along the Amazon, except Rio Puty (= Rio Potí) which is situated in the Brazilian federal state of Piauí (comp. *A. piauiensis*).

In a study from 1872, COPE describes *Geophagus amoenus* from ›River Ambyiacu‹ (= Rio Ampiyacú) in the Departement Loreto/Perú. The new species was first presented as related to GÜNTHER's *M. taeniatus*, but later it was synonymized with *M. taeniatus* by STEINDACHNER. Besides, the same author described another ›*Geophagus*‹ species: *G. badiipinnis*, which however later turned out to be a synonym of *Chaetobranchus flavescens* HECKEL. STEINDACHNER (1875) deprived the status of an independent genus from *Mesops*, considered it only a subgenus of *Geophagus* and listed it there together with *G. cupido*, the new species *G.* (*M.*) *thayeri* (= *Acarichthys heckelii* [MÜLLER & TROSCHEL, 1848]) and *G.* (*M.*) *agassizii*.

Also in works of EIGENMANN & EIGENMANN (1891), EIGENMANN & BRAY (1894) and BOULENGER (1895 [a + b]), the status of the subgenus *Mesops* was considered valid. BOULENGER described further species as ›*M. taeniatus*‹, which however later turned out to be a species that had to be described (*A. borellii*, *A. commbrae*, *A. trifasciata*).

In 1902, PELLEGRIN summarized his Cichlid list by using the collection compiled by

Jobert in Brasil in 1878. Among the species mentioned, also *G.* (*M.*) *taeniatus* was mentioned again. From todays point of view, there were many inaccuracies. Pellegrin's work however, does not only deserve particular attention because he was the first to remove the blinkers and started breaking the generic structure that was kept for a long time. He stated that the eyes' position -previously considered a characteristic trait for geophagus-like genera - should have less importance than the number of gill rakers and the position of the divided but often reduced, i.e. often only rudimentarily present lateral line.

In the years 1905 and 1906, Charles Tate Regan (1878-1943), the British Ichthyologist who was working at the British Museum in London from 1901 to 1938, composed the most extensive revision of American Cichlids - indeed with greater precision than Pellegrin. However, he did not completcly accept Pellegrin's opinions, in contrast to this, he formed a diagnosis that approached Pellegrin's, and for the Dwarf Cichlids he proposed the name *Heterogramma*, but once more without naming a generic type. Compared to *Geophagus*, the new genus was first characterized by the lower count of dorsal fin rays (5-7, *Geophagus* 9-14) and secondly by the upper lateral line, in many species approaching the base of the dorsal fin up to one scale row. When during his work with fish of the Rio Ucayali/Perú (1913) Regan discovered that the generic name *Heterogramma* was already given (›preoccupied‹) to another group of animals (Billberg, 1820 or Guenée, 1854), he proposed the name *Apistogramma*. Günther's *M. taeniatus* was adopted as generic type.

In 1954, Fowler - obviously without exact (!) knowledge of Regan's work - describes a new monotypic genus for *Apistogramma trifasciata* which was already described in 1903. As discriminating trait, the sawn anterior gill cover (praeoperculum) is given (page 316, 386). To the honor of Dr Pinto it is named *Pintoichthys*, a genus which nowadays has to be considered a younger synonym of *Apistogramma*.

Neither for the generic name *Heterogramma*, nor for *Apistogramma* did Regan mention the gender of the generic names. While subsequently handing in the new generic name, only ›*Hetero*-‹ was replaced by ›*Apisto*-‹. In translation, both names mean as much as ›irregular‹ or ›unreliable‹, -*gramma* (= line) means the lateral line which, as has already been said, is often reduced. So the author supposes that the lateral line is formed weakly respectively irregulary.

In his revision from 1980, Kullander deals in detail with the gender of the generic name. His remarks* reveal that the gender is not be seen as neutral, as it was supposed in the past, but as feminine. Knowing this is important for the suffixes of certain specific names which have to be adapted.

Meanwhile, with *A. ortmanni* Eigenmann (1912) added another species from British Guyana and eastern Venezuela to *A. steindachneri* which was described by Regan in 1908.

---

* »In not explaining explicitly the meaning of his names, Regan left us with a problem. There are two Greek words than can be latinised into *gramma*. The first is neuter and means letter, something written, or basic knowledge (cf. English words grammar, program), the second is feminine and means a stripe of or a line. If the meaning is considered, there can be no doubt about which word Regan had in his mind. Meinken (*in* Holly *et al.*) explains *Apistogramma* as meaning „mit unzuverlässiger Seitenlinie" (= with unreliable lateral line), i.e. the feminine word is one sought, and it retains in gender after latinization. -a is also the common Latin feminine ending. The "gender problem" was first observed by Schmettkamp (1976), who noted that neuter endings to specific names were commonest in literature, but that also masculine (*amoenus*) and, mistakingly, feminine (*pleurotaenia*) occured.«

In the following decades it happened repeatedly that Dwarf Cichlids were described under the name *A. taeniatum* by various scientists – however from far away areas where specimens of this species do not exist. So for example in 1922 EIGENMANN described the specimens from the upper Rio Meta in Colombia which were later described as *A. macmasteri* by KULLANDER. In the 1930s, the Berlin Ichthyologist Ernst AHL wrote descriptions of various *Apistogramma* species about which one later found out that as type material, only a few specimens were present (donations of ornamental fish importers with partial imprecise origin): *A. parva* (1931, »Estado do Pará, Brasil, Rio Capim«), *A. ornatipinnis* (1936 = *A. steindachneri*), *A. weisei* (1936 = *Taenicara candidi* MYERS, 1935), *A. aequipinnis* (1938, »supposedly Argentina«), and *A. reitzigi* (1939 = *A. borellii*).

Already in 1935, the American Ichthyologist George S. MYERS described *Taeniacara candidi* without picturing specimens, and therefore he created not only a new species, but also a new genus. The author classified the genus and the species as close to *Nannacara*, but today one tends rather to the opinion that *Taeniacara* has to be ranged in the affinity of the *Geophagini* and *Nannacara* (together with *Laetacara*) to the *Cichlasominii* (KULLANDER, 1983).

The description by HOEDEMAN (1951) shows how unclear the data about the origin of certain specimens was at that time and how little one could trust in the data. *A. cacatuoides* originates from HOEDEMAN's publications, which due to the given information should have been collected in Surinam, although later it was learned that these fish are not found there, but selectively (in a wider range around the city of Iquitos) in the upper Amazon (Ucayali drainage).

Between 1960 and 1971, Hermann MEINKEN (1896-1976) who is still known to most aquarists as leader of the VDA Fish Definition Department, described a number of *Apistogramma* species - partially even with single specimens - whose origin could not always be considered assured. In 1962, the author presented a new species and started the »attempt of a survey of the genus«. He accepted 25 species, but he overlooked *A. bitaeniata* - and this was to have consequences. Already in 1961, he described *A. sweglesi* (»near Letitia in Peru«, today it is considered an ›uncertain species‹ and probably has to be seen as a synonym of *A. bitaeniata*) and *A. klausewitzi* (= *A. bitaeniata*). In 1964, the description of *A. kleei* (= *A. bitaeniata*) followed, in 1965 the first description of an *Apistogramma* species from Venezuela which today is still considered valid: *A. hoignei*. In 1969, *A. gibbiceps* was described, in 1971 *A. geisleri* from the surroundings of Óbidos near the mouth of the Rio Trombetas into the Amazone River and *A. borellii* (from the surroundings of Manáus) which was later described by KULLANDER as *A. regani*.

Apistogramma pucallpaensis was considered close to *Apistogramma* by MEINKEN with justification (as is expressed by the generic name!). In this concern, the author found out two obvious distinguishing traits: In the anal fin there are eight hard fin rays. As a rule, in *Apistogramma* species there are only three, rarely four. Besides this, the anterior lateral line which is situated above, is always further away from the base of the dorsal fin than one scale width.

In one of his first works on *Apistogramma* in 1976, KULLANDER described *A. luelingi*. The fish named after the German Ichthyologist Karl-Heinz LÜLING was imported by him from the north of Bolivia (drainage of the Rio Chaparé/Rio Madre de Dios) already ten years ago, whereby 15 specimens, not completely adult, were caught, preserved and brought to Germany. First they were given to MEINKEN who, however assigned them to the species *A. cacatuoides* which at that time was incorrectly known as ›*A. borellii*‹. KULLANDER later saw the differences. In fact, both species show certain family traits, but the anal fin of *A. luelingi* shows more hard rays than it is generally common in *Apistogramma*. With the description of four new species from the system of the upper Rio Orinoco (*A. macmasteri*, *A. hongsloi*, *A. viejita* and *A. iniridae*), in 1979 KULLANDER continued his work on *Apistogramma* and in a revision (1980) as well as later he described further new species.

It must not be overlooked that in 1977 KULLANDER created a new genus by describing *Papiliochromis* for *A. ramirezi*, to which in 1981 the Swedish Ichthyologist also classified *Crenicara altispinosa* HASEMAN, 1911.

KULLANDER's diagnosis of *Apistogramma* (1980)

= the comprehensive description of the most important traits for the systematic position of the genus in the system.

The following four traits are listed as particularly distinctive:

- gill rakers on the sides of the lower plate of the pharynx teeth (=Pharyngeals).
- compressed lobes on the first lateral gill branch (Epibranchial) with gill rakers on the edge.
- three, more rarely four or six, anal spines.
- 14 to 18 dorsal spines.

These traits discriminate the members of the genus *Apistogramma* from:

- *Apistogrammoides* MEINKEN, with the only species (of the genus) which shows 7 to 9 anal spines, but which has an equally different course of the upper lateral line, with a different colour pattern and so on.
- *Geophagus* HECKEL, the presently 10 species whose members have no gill rakers on the tooth plates but which are also furnished with another colour pattern, with deeper preorbital bones and a deeper lacrimal bone, with teeth on the fourth gill arch (Ceratobranchial), more gill rakers, with the course of the lateral line and so on.
- *Biotodoma* EIGENMANN & KENNEDY, the representatives of the presently two species have no gill rakers on the toothplates, but they also differ by another colour pattern, a deeper preorbital bone as well as two specified nerve headbones (Supraneural*) instead of one and so on.
- *Gymnogeophagus* A. DE MIRANDA RIBEIRO. The presently four species have no gill rakers on the toothplates, but they have a different colour pattern, no supraneural and so on.
- *Papiliochromis* KULLANDER (today *Microgeophagus* FREY [the authors]). Their representatives have no gill rakers on the toothplates, but a different colour pattern. The lateral line branches off towards the caudal fin. The specimens have a different tooth order, less dorsal spines and so on.
- *Acarichthys* EIGENMANN. The two representatives (*A. geayi* was meanwhile classified into another genus [the authors]) have no gill rakers on the toothplate, only one of them (*A. heckelii* [the authors]) has a compressed lobe on the lateral gill arch, a different colour pattern, a different course of the lateral line and so on.
- "*Crenicara*" *altispinosa* HASEMAN. Obviously, the species has no epibranchial lobes, but it also has a different colour pattern and so on. The species was meanwhile classified through *Papiliochromis* into *Microgeophagus* (the authors).

---

* Supraneurals are unpaired nerve spines of the anterior vertebra between the skull and the anterior dorsal spines. They are also called ›predorsal bones‹.

- *Biotoecus* EIGENMANN & KENNEDY. The presently only species (meanwhile another one has been described [the authors]) has seven or eight dorsal spines.
- *Taeniacara* MYERS. The only species having no epibranchial lobe and a lateral line is missing.

Synonyms

Younger synonyms are as follows:

Heterogramma corumbae EIGENMANN & WARD = *A. commbrae.*

Apistogramma ornatipinnis AHL = *A. steindachneri.*

Apistogramma reitzigi AHL = *A. borellii.*

Apistogramma amblopIitoides FOWLER = *Acaronia nassa.*

Apistogramma wickleri MEINKEN = *A. steindachneri.*

Apistogramma trifasciatum haraldschultzi MEINKEN = *A. trifasciata maciliensis* (the authors).

Apistogramma klausewitzi MEINKEN = *A. bitaeniata.*

Apistogramma kleei MEINKEN = *A. bitaeniata.*

or synonyms due to thorough examination of the original description:

Heterogramma ritense HASEMAN = *A. borellii.*

Apistogramma weisei AHL = *Taeniacara candidi.*

or considered invalid:

Heterogramma ortmanni EIGENMANN in HASEMAN.

Apistogramma taeniatum longirostris MEINKEN.

Physiognomy, colour patterns and fin forms

Knowing the name of an *Apistogramma* species is one side of the coin. Trying to find the specific name of an unknown wild fish only depending on its origin, is not enough. Some of the discriminating traits mentioned in KULLANDER's scientific work can hardly be verified by most aquarists. This is not true for those defined by physiognomy or those belonging to the basic colour patterns - even if these more or less depend on the fish's mood. Many Dwarf Cichlid keepers should already know that most of the various *Apistogramma* species (and here mainly the males) can also be discriminated by the form of their caudal fins.

One talks about a morphe and thereby means the shape, body form or the appearance of the specimens. Correspondingly, morphology means the science of shape and body form of animals and morphometry stands for the measurement of the body.

The head is measured from the tip of the snout to the bow of the posterior gill cover (GC). The trunk follows and ends at the anus (TR). Then follows the caudal length reaching to the beginning of the caudal fin (CL). The caudal peduncle is included (CP). It also ends at the beginning of the caudal fin, but it starts at the beginning of the last fin ray of the anal fin. In scientific publications, mostly only the standard length is given (SL). It differs from the total length (TL): whereas the former reaches only from the tip of the snout to the end of the caudal peduncle, the latter includes the caudal fin and therefore comprises the length of the fish ›totally‹.

The body deepness (BD) is measured at the deepest part of the body. A more detailed explanation of the habitus of a Dwarf Cichlid requires also data about the body shape because this character is not united in all members of the genus *Apistogramma*. They can be relatively easily distinguished as ›deep bodied‹, ›slender‹ or ›elongated‹.

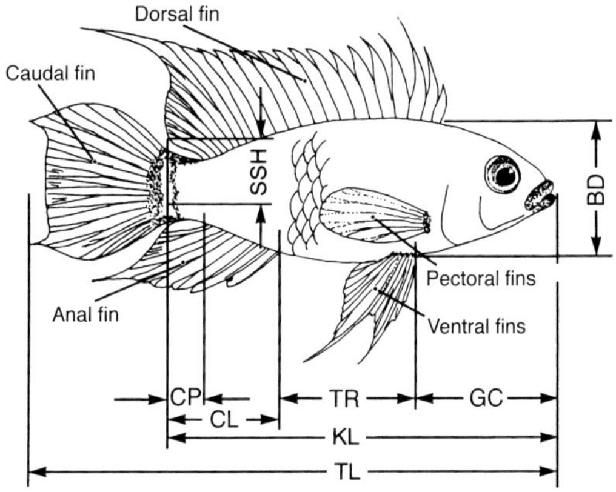

Cichlids have a continuous dorsal fin which is segmented in a (anterior) part with hard fin rays (= spines) and a posterior one with soft fin rays. Then there are one pectoral

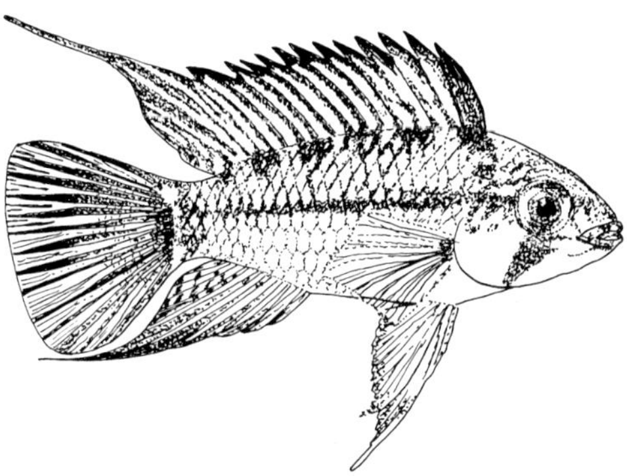

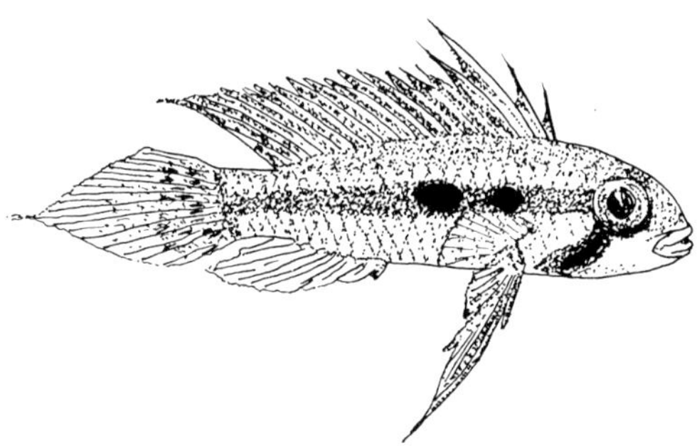

and ventral fin on each side, as well as the already mentioned caudal fin. Particularly in males, the differences in the form of the dorsal and caudal fin attract attention. Concerning the former fin, one distinguishes between a relatively flat lying fin from which the membranes stand out more (sawn) or less (truncated). Doubtlessly, among the more attractive specimens are those with sawn membranes and anterior fin rays that tower high above the rest of the fin. Finally there is the dorsal fins whose spines and membranes are extended and whose towering membranes are grown together forming a sail-like shape.

Specimens of both sexes often differ considerably in colouration and body size. Generally, males show a more magnificent finnage whose particular - specific - traits suffice for distinction.

The forms of the caudal fin characterize themselves by the growth of the fin rays and membranes. In the border areas they can be truncated or sawn, the membranes of the anterior spines can be more or less extended or the membranes of the extended spines grow together near the upper fin edge, giving the dorsal fin a sail-like appearence.

The forms of the caudal fins are also a good help for determining the affiliation to a group of species to an *Apistogramma* species.

There is a differentiation between rounded, oval or truncated form as well as the particularly conspicuous lancet-like form in *A. agassizii* and another form whose ends are bilobated and partially provided with strong, filamentous extensions. With corresponding guidelines, many *Apistogramma* species can already be classified with these traits.

In many specimens, the basic colour pattern (black markings) often appears depending only on the mood (affection, stress and others). But also in preserved specimens it can also be recognized over a long period of time. The head is coated with several bands. From the eye to the neck there is the superorbital or forehead stripe. It is followed by the suborbital or cheek stripe. From the upper lip through the eye, up to the end of the caudal peduncle (occasionally also into the caudal fin) there is a stripe which is subdivided into two parts. Its anterior part in front of the eye is called preorbital or snout stripe, the part running across the flanks is called lateral band. In the center of the lateral band, there can be one or more black blotches. In this case one speaks about the lateral blotch/blotches. The lateral band ends mostly at the posterior caudal peduncle, with a round, angular or vertically oval caudal blotch. Those bands running vertically across the flanks in a straight or slightly convoluted way are called transversal bands. They start mostly in the dorsal blotches, but can also stretch into the base of the dorsal fin. The ventral part of the body can show an additional pattern. Horizontal lines are called abdominal or ventral lines. If the lower edge of the ventral part is seamed in black, one speaks of a ventral band. If the anal zone is surrounded by a black blotch, this one is called anal blotch. If the base of the pectoral has a black marking, we call this a pectoral blotch.

Family relations
Division into species complexes

The growing number of species of the genus *Apistogramma* caused KULLANDER to sum-marize the, by then described species, in complexes of a certain group of associated species in his revision of 1980, as this was already done by some authors before him (MEINKEN, 1962; GOLDSTEIN, 1970). To the Swedish author, the result of these works seemed to be too artificial, too highhanded and therefore useless, so he formed a new division into groups, respectively complexes, which here was completed by some later descriptions of species. STAECK (1991[c]) defined the term ›species complex‹ as follows: »Contrary to a species group which possibly is polyphyletic*, by a species complex I understand a monophyletic* group of associated species.«

The *regani* complex comprises specimens with clearly pronounced vertical, horizontal and/or head bands. An anterior dorsal blotch is present, whereas a blotch on the flanks is absent. There are no sexually conditioned differences in the dorsals and caudals. Relatively deep bodied shape.

Today in this complex species like *A. caetei*, *A. cruzi*, *A. eunotus*, *A. geisleri*, *A. gossei*, *A. moae*, *A. ortmanni*, *A. pleurotaenia*, *A. regani*, *A.resticulosa*, *A. urteagai* and (possibly) *A. piauiensis* are included.

The *macmasteri* complex is considered by KULLANDER the most regular one. All species comprised here live in the Orinoco basin. *A. hoignei* is distributed into the delta. For these specimens, a dark pectoral blotch in the females can be seen as a particular trait. The representatives of these species are also moderately deep bodied. Vertical bands are well pronounced; a possible lateral horizontal band consists of a band of dark spots. Head stripes and anterior dorsal fin rays are black. In the males, the front of the dorsal has extended membranes. The caudal is rounded.

Today, in this complex species like *A. macmasteri*, *A. viejita*, *A. hongsloi*, *A. guttata* and *A. hoignei* are included.

The *commbrae* complex is distributed in the southern Mato Grosso (Corumbá) and in the neighboring Paraguay. Therefore, KULLANDER also speaks about »the Paraguayan counterpart« of the *regani* complex. The moderately deep bodied Dwarf Cichlids show several vertical, and one horizontal band. The anterior dorsal membranes show a dark tinge but no pectoral blotch. There is no sexual difference in dorsal and caudal.

Today, species like *A. commbrae*, *A. inconspicua*, *A. linkei* and (possibly) *A. aequipinnis* are placed in this complex.

The *borellii* complex and the few species named here do not necessarily let us see a

* monophyletic = deriving from one ancestral form; polyphyletic = relating to phylogenesis, deriving from several ancestral forms

family correlation. Wild specimens are known from all species. The body is quite deep. In the males, dorsal, anal and pectoral fin edges are conspicuously extended.

Today, some authors place representatives of the aquaristically well known species *A. borellii* as well as *A. rondoni* (de Miranda Ribeiro, 1918), described from the Rio Paraguay (Cáceres/Mato Grosso) into this complex, whereas the latter possibly has to be considered a synonym of the former. Moreover, *A. brevis*, *A. staecki*, *A. maciliensis* and *A. trifasciata* should be included.

The *agassizii* complex is represented by at least three interesting species. Two from newer descriptions could be added. These are specimens with an elongated body shape and with only faintly conditioned vertical and horizontal bands as well as the lateral blotch. A pectoral blotch is absent.

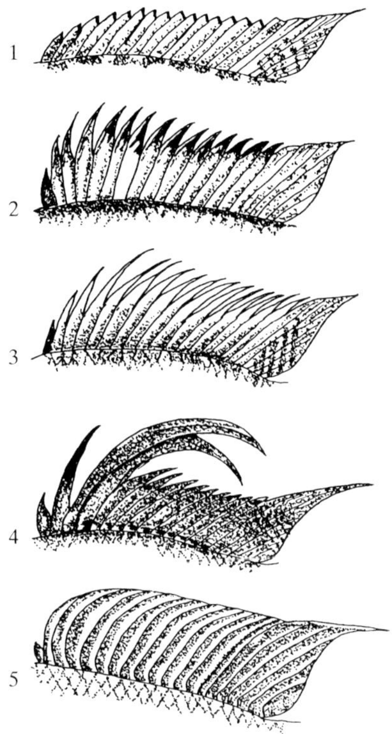

1 The truncated membranes extend only slightly over the spines
2 The sawn membranes extend here also slightly over the spines
3 Here the sawn membranes are further extended over the spines
4 Membranes are sawn and in the anterior part of the fin extend over the spines
5 So-called sailfin in which the extend spines have extended and fused membranes

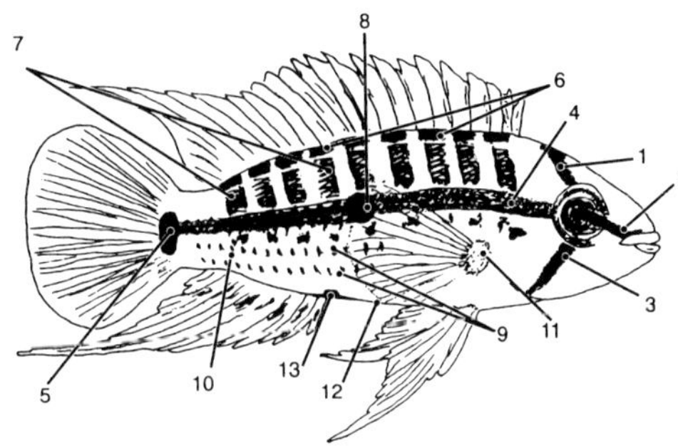

1 Forehead or superorbital stripes
2 Snout or preorbital stripes
3 Cheek or suborbital stripes
4 Lateral band
5 Caudal blotch
6 Dorsal blotches
7 Transverse or vertical bands
8 Lateral blotch
9 Abdominal stripes
10 Abdominal strokes
11 Pectoral blotch
12 Ventral band
13 Anal blotch

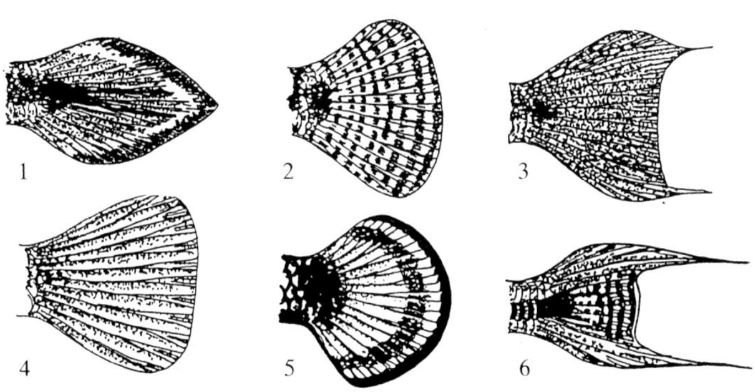

1 Lancet-like form
2 Rounded form
3 Bilobated form
4 Truncated form
5 Rounded form
6 Bilobated form with extended filaments

Dorsal fin with one or more extended membranes. Anterior fin rays in the males intensely extended and at the end (together with the anal fin) extended with tips. Central caudal fin rays are extended, by which the caudal fin receives an arrow-like appearance. The caudal can be bilobated or rounded and then in the former case it has an extended fin ray in each lobe.

In this complex species like *A. agassizii*, *A. bitaeniata*, *A. elizabethae*, *A. gephyra* and *A. paucisquamis* are placed.

The *gibbiceps* complex unites specimens that have a moderately elongated body as adults. The body pattern is mostly expressed by slopely arranged vertical bands. Additionally, an emotionally intensely shown horizontal band can be found. In males, the caudal fin is bilobated.

Today, in this complex *A. gibbiceps* as well as (possibly) *A. personata* are placed.

The *pertensis* complex with the conspicuously slender representatives can be distinguished best of all. The males of three species (described until 1980) have - at least in the posterior dorsal fin - membranes which grow together beyond the tips of the hard fin rays. This gives a sail-like appearance to the quite high fin. In the males, rounded or lancett like caudal fins are also known. Vertical bands may be recognizable, but they can not be called part of the body pattern. A lateral horizontal band as well as a lateral blotch can be seen. Head bands are present only incompletely and a black colouration of the anterior dorsal fin rays cannot be detected.

Today, species like *A. pertensis*, *A. iniridae*, *A. meinkeni*, *A. pulchra* and *A. uaupesi* are placed in this complex.

In the description of *A. hippolytae* (1982), KULLANDER created the steindachneri complex and among others he determined the following traits: a »big obvious caudal blotch; striped caudal fin; black anterior dorsal fin membranes; well pronounced head stripes; completely scaled pectoral area; absence of a pectoral or caudal blotch; no extended dorsal fin in the males, but caudal round to slightly truncated (*A. hippolytae*), slightly truncated (*A. rupununi*) or with dorsal and ventral extensions (*A. steindachneri*).«

A. steindachneri, *A. hippolytae* and *A. rupununi* are in this complex.

Besides the species comprised in complexes, there are others which following KULLANDER's opinion of that time could not yet be classified in one of the given complexes. Today, these are comprised in a *cacatuoides* complex.

Here, besides the polymorphic *A. cacatuoides* also *A. norberti*, *A. luelingi* and *A. juruensis* are included. Possibly *A. nijsseni* and *A. payaminonis* belong to this complex. Assuming from the described classification into the *cacatuoides* complex, it has to be stated that in it, two species groups can be detected whose distribution scheme (at least following our present knowledge) seems to be intensely limited (*A. nijsseni*, *A. payaminonis*), on the other hand it seems to be widely extended (*A. cacatuoides*, *A. luelingi*, *A. juruensis*).

As the habitats of *A. payaminonis* (in the system of the Rio Napo [Ecuador/Perú]) and *A. juruensis* (in the system of the Rio Juruá [Brazil]) until today could not be examined in a sufficiently detailed way, we have to wait and see what future research will show.

Reproductive Behaviour

Despite their wide distribution, the reproductive habits of representatives of *Apistogramma* species is quite unified. They spawn in hiding places or caves with strong territorial behaviour, during the spawning season, so that in the natural habitat the males gather several females around them and thus are polygamous (= living in intermarriage).

Besides, one speaks about a conspicuous sexual dimorphism, the occurrence of two differently configured forms in the sexes of a species, whereby here the females always are the smaller specimens. By this, polygamy and sexual dimorphism have a certain dependence to each other.

Raising the fry happens in a so called, male-mother-family, in which the immediate work with the young fish is done by the mother alone, whereas the father cares for the defence of the territory in the periphery of the breeding zone.

Occasionally, one can also observe pairs in which the male is endeavoured to contribute his part in leading the offspring. If in a breeding aquarium of adequate size several females are kept, each of them forms its own limited breeding territory separately. The added male has to be a strong "self confident" specimen.

After the female has deposited the eggs at the ceiling of the cave, these are fertilized by its partner.

This is not always easy as the female often adapts the dimensions of the selected cave to its own smaller body measurements and then the cave can be so narrow that in fact the bigger male does not fit, or only partially. Therefore, a cave installed by the aquarist should indeed be narrow, but not too narrow. Then the female itself narrows the entry, as this corresponds to its need for safety.

About the colouration of the specimens

The pigment cells are responsible for the colouration of the fish. They contain various dyes after which they are classified. Melanophores contain melanine, a black-brownish granulous hence solid dye. The guanophores or iridocytes also contain a solid but crystalline dye called guanine respectively pterine, which also belongs to the guanines. To many fish bodies they confer a silvery colouration. Chromatophores mainly contain liquid carotines and flavines. The former ones are yellow, orange or red and distinguished by their solubility in fat. Flavines are yellow, not soluble in fat, but strongly fluorescent. Whereas iridocytes mostly contain a fix pigment, the contents of melanophores and chromatophores can be transferred to the cellular centre by sympathic (concerning the sympathicus, sensitive to a part of the vegetative nervous system) nervous stimulation or by hormonal control (colour centering) or spread all over the cell (colour dispersion). These nervous mechanisms cause the often surprising colour changes of some fishes. The different colours and mixed tones are caused by the combinations of various chromatophores and melanophores above the light reflecting guanine layer.

In Dwarf Cichlids, colour changes often occur in dependance of the emotion. Black patterns are preferrably shown under stress. However, the specimens show their most magnificent colours while mating, spawning and fertilizing as well as while caring for the fry. Females show a particular breeding colouration. During the breeding season, many *Apistogramma* females adopt a lemon to golden yellow colouration which is accompanied by an intense black colouration of the anterior dorsal spines and partly also of the anterior rays of the ventral fins. In females of *Microgeophagus ramirezi* a red colouration in the ventral zone is conspicuous, this is known from Dwarf Cichlids of the West African genus *Pelvicachromis* in a similar way.

About the special colouration of the females

Buying tropical fish is said to be a ›confidential matter‹. This would also comprise the fact that the shop assistant or breeder gives his customer a female of the same species for the selected *Apistogramma* male. Normally, this is the case, but perhaps it is only possible because the specimens swim in sale tanks ordered in species. If females of many different species would be together in one tank, there would be certainly errors in identification.

»For me, all females look approximately identical!«, assured me once a Dwarf Cichlid keeper who had already been able to reproduce several offspring. Dear reader, do you believe you could assign infallibly the specifically suitable female to every known male? Perhaps it might work, but there are borderline cases in which the females of different species are very very similar. In this concern it has to be taken into consideration that while being caught out of the salesman's or breeder's tank, besides their more or less known standard and breeding colouration, the females can show an additional shock colouration which clearly differs from those mentioned before and show many more dark tones.

What can happen if the wrong female is given to the male of a desired species? The aspired ›marriage‹ firstly cannot come into being for reasons of wrong partnership. But secondly, under certain conditions it can happen still, and thereby it can bring offspring to the breeder. RÖMER (1995) and also KOSLOWSKI (1995) have repeatedly treated this aspect in their reports. As an example the form ›orange fin‹ is named. In this concern, RÖMER remarks: »various parental combinations resulted in products that are extremely similar to the orange fin *Apistogramma*. The most important observation made during the experiments is that almost all hybrid combinations lead to offspring with scale anomalies. Of course it has to be asked why in crossings of *Apistogramma* species the depicted appearances occur. The cause is to be seen in a compatibility of genetic information of both parents that is only incomplete.« Of course, scale anomalies can have other causes either, and the mentioned author refers as follows: »... for example serious fighting injuries occasionally lead to abnormal reformation in the range of the damaged tissue. Very bad water conditions, but first of all high nitrate figures can partially lead to extensive deformations of fins and scales, too.«

It should be the duty of every aquarist seriously working on the reproduction of Dwarf Cichlids to get a precise idea of the appearance of a female fitting to a male of the same species. KOSLWOSKI (1995) adds figures of 16 females to his report and divides the females into 12 groups with different combinations of traits.

The statements about the looks of the females given in this book can only be based on living specimens. Contrary to this, in KOSLOWSKI (1995) »...for their identification besides the body shape and markings on the caudal fin traits are also used that can only be verified at preserved material (for example the arrangement of head pores, compare KOSLOWSKI 1985[a]) and which have therefore only minor value for an aquarist or other viewers of living specimens.«

Standard and mating respective breeding colourations differ from each other. Above all, the latter is distinguished in most females by its intense yellow colouration as well as by a stronger black colouration which is thus fuller in contrast to which is referred to - at least in every scientifically described species. During the breeding season the black markings are quite constant. However, they are almost always shown only by the ›right‹ female, which means by the specimen that has already started a ›marriage‹ with its partner. But as the black markings can not only vary outside of the breeding season, but also has to be seen in coherence with the again intensely differing shock colouration, the aquarist who is not certain while watching a female specimen, should search for advice in information given in various kinds of literature as well as by *Apistogramma* specialists.

The right water: Condition for a successful reproduction

As already mentioned, the water in most of the distribution areas of South American Dwarf Cichlids is very soft and more or less acidic (cp. also the various water analyses). Nitrate comes out of the tap,with an unpleasant concentration at many places - combined with traces of fungicides, herbicides and pesticides. Nitrate is absent in tropical waters or present only in faint concentration. As most of the aquarists only concentrate on the measurement of the total hardness, it shall be said here that in fact the carbonate hardness is mostly included in the total hardness (name!), but this does not have to be true: the carbonate hardness can exceed the total hardness. But we find this combination less in South American than in African waters (for example in the Lake Tanganjika). Even if perhaps more expensive in acquisition of devices, electronic measurement of the present mineral combination in Mikrosiemens (µS) is easier to survey than the well known method with liquid colour indicators. Here it would lead too far to present detailed information about handling the preparation of soft water. This has already happened in many of our own publications as well as in those of other authors (cp. MAYLAND, various Discus books in the bibliography). Besides, in areas with only few industries one can refer to rain water. In order to keep the nitrate value low, it is additionally recommended to make partial water changes as often as possible (in a 3 to 4 day rhythm).

Among the Dwarf Cichlids presented here, there are some which place minor demands on the water quality. To these belong *Apistogramma borellii*, *A. cacatuoides*, *A. commbrae*, *A. guttata*, *A. hoignei*, *A. linkei*, *A. luelingi* and *A. resticulosa* as well as - conditionally - *A. hongsloi*. The water does not have to be extremely soft, but it should not exceed 4 to 6 degrees of total German Hardness respectively 200 µS. The pH value should be in the neutral range, which means at ± 7.0 (if possible below that value, but not above). Similar conditions are also required for the two substrate spawning representatives of the genus *Microgeophagus* (formerly *Papiliochromis*), whereby for *M. ramirezi* it seems to be more recommendable to choose softer water with a slightly acidic pH value.

Contrary to the previously mentioned species, Cichlids from the central Amazon, the Guayana countries, or parts of the Orinoco drainage, are more sensitive. They require very soft water and a pH value that is definitely in the acidic range. This is true for species like *Apistogramma agassizii*, *A. bitaeniata*, *A. elizabethae*, *A. gephyra*, *A. gibbiceps*, *A. gossei*, *A. nijsseni*, *A. norberti*, *A. iniridae*, *A. macmasteri*, *A. paucisquamis*, *A. personata* and *A. pertensis* as well as others. Similar conditions are required by substrate breeding species of the genus *Dicrossus*.

When one is thinking about buying a smaller device for reverse osmosis - influenced by hints in Discus books - one should consider that Discus are much bigger, and due to their regular water changes, a comparably much larger volume of water is needed. Besides, the acquisition of such a device is price related because only good and thus more expensive devices have an acceptable gain of pure water. For inferior devices the gain of pure water is only 20%, whereas 80% is considered as waste water. The pure water obtained this way has not only lost all the hardening minerals, but also nitrate and the possibly contained fungicides, herbicides and pesticides that inhibit the development of the eggs and can not be used in this form. It is dangerous to life, therefore minerals and trace elements - even in low concentration - have to be added. For this purpose, aquaristic trade mostly offers corresponding mixtures. Adding once more tap water to the aquarium water would be senseless because its damaging components had just been deprived beforehand by an expensive method.

Biotope: Venezuela, Rio Jaspe, north of the town of Santa Élena de Uairen, where in the south-eastern corner of the Gran Sabana for several kilometers the river flows over hard, rusty red, Jaspis ground, and in this area, it also flows over a medium high waterfall.
Date: 12.02.1991
Time: 16.00 o'clock
Water movement: Moderately flowing
Water depth: Depending on the width of the river bed, however on the average it is shallow
Water colour: Clear

Laboratory results:

pH-value	6.91
Electrical conductivity (25°C), µS/cm	11
Acid capacity up to pH 8.2 (p-value), mmol/l	<0.05
Acid capacity up to pH 4.3 (m-value), mmol/l	0.12
As carbonate hardness, °d	0.34
Total Alkalis (Ca^{2+}/Mg^{2+}), mmol/l	0.005
As total hardness, °d	0.03
Sodium (Na^+), mg/l	1.0
Potassium (K^+), mg/l	1.1
Calcium (Ca^{2+}), mg/l	1.2
Magnesium (Mg^{2+}), mg/l	<0.1
Total Iron ($Fe^{2+/3+}$), mg/l	0.20
Manganese (Mn^{2+}), mg/l	0.002
Ammonium (NH_4^+), mg/l	<0.02
Nitrite (NO_2^-), mg/l	<0.02
Nitrate (NO_3^-), mg/l	<0.1
Chloride (Cl^-), mg/l	0.13
Sulphate (SO_4^{2-}), mg/l	<0.1
Total Phosphate (PO_4^{3-}), mg/l	<0.04
Hydrogen Carbonate (HCO_3^-), mg/l	7
Oxidation	
a) As Potassium Permanganate ($KMnO_4$) consume, mg/l	5.5
b) As Oxygen (O_2), mg/l	1.4

Mutations, mutagenes, mutants and pedigrees

A mutation is a change of the genetic material that occurs either spontaneously, or is caused artificially by mutagenes like radioactive radiation. Contrary to environmentally and thus mostly only temporarily caused inhereditary deviations in the external appearance (for example in hunger forms), these are permanent and therefore hereditary changes. These erratic changes (lat. = *mutatio*) can be easily detected in so-called mutants as aberrant single specimens in the entirety of a group of offspring.

Mutagenes are artificial aids for reaching a change (which is then not spontaneous any more) in order to enhance the mutation rate. For an artificial ignition of the gene or chromosome mutation, in a mostly experimental way one uses chemicals or ionising radiation (gamma, Röntgen or UV rays, neutrons). As we know from reports in media, such experiments are mostly more harmful to the genotype than beneficial, and the number of animals coming out of such experiments as cripples and having to be killed immediately, can be high.

Assuming the fact that the specimens of many *Apistogramma*, *Crenicara* or *Dicrossus* forms which are used today for breeding further generations descend from wild fish, accordingly we deal with normal offspring from wild fish of the corresponding species! If there is a mutant occuring in species of these genera that differs obviously from the known standard to be expected from the offspring, it is often only a single specimen. Then the attempt to backcross can show if the mutant deserves its name.

For animals of healthy breeding forms, the follwing principle is effective: »The better, and in the male form, the more colourful the physical general impression with erected fins, the more desired is the specimen!«

Biotope: Venezuela, Rio Orinoco, west (upstream) the city of Ciudad Bolívar. In the upper part the coffee brown water is strongly mixed by various tributaries rich in sediments, and with that it has lost its former colour
Date: 15.02.1991
Time: 14.30 o'clock
Water movement: Rapidly flowing
Water depth: Not established
Water colour: Brownish, clouded by sediments

Laboratory results:

pH-value	6.90
Electrical conductivity (25°C), µS/cm	44
Acid capacity up to pH 8.2 (p-value), mmol/l	<0.05
Acid capacity up to pH 4.3 (m-value), mmol/l	0.35
As carbonate hardness, °d	1.0
Total Alkalis (Ca^{2+}/Mg^{2+}), mmol/l	0.17
As total hardness, °d	0.9
Sodium (Na^+), mg/l	2.1
Potassium (K^+), mg/l	0.5
Calcium (Ca^{2+}), mg/l	4.8
Magnesium (Mg^{2+}), mg/l	1.1
Total Iron ($Fe^{2+/3+}$), mg/l	0.51
Manganese (Mn^{2+}), mg/l	0.017
Ammonium (NH_4^+), mg/l	0.02
Nitrite (NO_2^-), mg/l	<0.02
Nitrate (NO_3^-), mg/l	<0.1
Chloride (Cl^-), mg/l	0.55
Sulphate (SO_4^{2-}), mg/l	3.3
Total Phosphate (PO_4^{3-}), mg/l	0.06
Hydrogen Carbonate (HCO_3^-), mg/l	21
Oxidation	
a) As Potassium Permanganate ($KMnO_4$) consume, mg/l	18
b) As Oxygen (O_2), mg/l	4.6

Such a case of ›assistance to nature‹ can not be equated with evolution. In the latter, at least in most cases, there is a gradual change in which the course of phylogeny through a progressing procedure has worked its way up to the highly organised forms of today. Here also the incidents of mutation play a role.

The selective theory knows four main factors of evolution. In certain populations they become effective and find their reference also in aquaristic publications:

1. mutation and recombination
2. selection
3. random effectivity
4. isolation.

Mutation was already treated in preceding smaller sections. Recombination means the total number of processes in whose course there is the formation of a new genetic combination. The concerned genes can be localized on different chromosomes.

Selection (DARWIN founded the so-called theory of selection, referred to natural selection in nature) is a procedure with which the most magnificent and the strongest specimens of a population reproduce with each other in a natural way (for example hierarchy) or wished by man (for example in the breeding of ornamental fish).

Breeders of *Tropheus* varieties from Lake Tanganjika, for example, know the effectiveness of living in isolation or of other environmental dependencies, it can also occur in the peripheral areas of South American waters: If a Cichlid population is isolated in its territory, and for example cannot cross a contiguous zone of deep water where big predators are waiting, it produces offspring that is typical for this population. In the course of the further development, through many generations, traits for a new species can be found.

Mutation and genetic doctrines presuppose a mathematical way of thinking. In recent times, concerning ›mutation‹ there were supporting as well as contrary arguments.

Clearly, the natural selection was the most important keynote within the evolutionary theory, whereas mutation was counted only as a game of nature. But research has shown that there could not always be the ›traditional paths‹ and that mutation had to be involved in order to create new possibilities where the nature of our earth was subject to permanent changes. In the past, ice ages were repeatedly replaced by droughts, liftings and lowerings of the earth's surface followed and without the ›intervening potential‹ of mutation, many steps would have taken too much time.

Once, in a work, the well known genetic researcher and mathematician R.A. FISHER remarked: »It consists of never letting the supply of genetic changes become scarce.« The experiments in many laboratories show that most of the mutations are harmful and commonly the most dramatic ones are even lethal. They have a negative feedback effect in a way that every change in a harmonic well adapted organism can have disadvantages. So many carriers of profound mutations do not stay alive long enough to bequest the changes to their offspring. Slight mutations - so they declare - have a higher chance of survival.

As far as we can observe, the occurence of changed trait in Dwarf Cichlids (and here mainly in the genera *Apistogramma* and *Microgeophagus*) has fortunately been restricted to patterns or colouration. Let us hope that there is not more and that the ›mayflies‹ - those conspicuous varieties that exist only in one specimen - are healthy too ... and remain so.

Whether mutation or pedigree, in South American Dwarf Cichlids, some became recognised. For example, we know the form of the Golden Ramirezi, a veil form of *Microgeophagus altispinosus* (which contrary to the former one obviously could not put itself through). Also in representatives of the genus *Apistogramma* some breeding forms can be listed, whereby the ›Rotschwanz-Agassizii‹ has to be named as well as the various states in the development of *A. cacatuoides* (›Zweifachrot‹ and ›Dreifachrot‹) or the forms of *A. hongsloi*, of which there are magnificent specimens.

Food and feeding

In their natural habitat, depending on the species, Dwarf Cichlids feed on the various offerings which nature holds ready in the corresponding season of the year. Even when in the aquarium, Cichlids prefer meat, therefore they can not be designated markedly predatory - rather can they be seen as omnivorous with a preference for meat. In this concern, it should not be taken amiss if they occasionally consider a young fish which fits into their mouth as welcome prey.

Customarily, Dwarf Cichlids take their food from the ground, whereby some of them show a particular jerky way of swimming. As most fishes, they also love a diversified diet which unfortunately is not always offered to them.

As most of the Dwarf Cichlid amateurs can be considered advanced aquarists, the actual feeding of the fish should be no subject. But as many aquarists particularly cherish the red tones of their fish, it has to be said that stronger red colours, as especially known in some breeding forms (for example *Apistogramma agassizii* "Rotschwanz", *A. cacatuoides* "Dreifachrot", *A. hongsloi* "Farbform II", cp. pictures), can not be permanently maintained brightly with flake food, *Enchytraea* or deep frozen red moskito larvae. Red colours have to be supported by a corresponding nutrition. Fresh (which means living) *Crustaceae* like *Daphnia* that can be caught in ponds are particularly qualified.

In Western Europe alone, there are more than 20 species. Furthermore, together with *Daphnia*, the collector mostly captures *Cyclops* of which *Cyclops fuscus* is the most frequent one.

As the members of both *Crustaceae* groups also eat microscopically small algae, the fish eating these *Crustaceae* simultaneously take a lot of vegetable components. Beyond this, the chitine armour of the *Crustaceae* forms a good fibroid food supporting the alimentary organs. Occasionally, one says that *Cyclops* have better nutrition values, besides this, these *Crustaceae* can be captured well during almost all seasons, whereas *Daphnia* can be captured mainly during the summer months.

The nauplii (larvae) of *Cyclops* can occur in masses even in early spring. Indeed, the aquarist has to dose the amount of the larvae because - if not eaten in time - they grow faster than the fry and can then endanger the small Cichlids.

Also for other reasons live food like this should be given within a certain time. They need quite a lot of oxygen and can suffocate if the container they are kept in is not sufficiently aerated.

In breeding Dwarf Cichlids, after the larval stadium the young fish are mostly developed far enough so that they can eat newly hatched brine shrimp nauplii. It is possible to buy the ready-to-hatch eggs of these *Crustaceae* in almost every pet shop. For hatching, a reasonably priced, so-called incubator, made of plastic should be purchased and with this, a water bottle is used. For this kind of live food it is also true that it should be given immediately after hatching (24 to 36 hours after starting the setup) because otherwise they quickly lose energy by their own swimming activities and therefore lose nutritive value.

Some of the animal pigments the fish owe their colours to, belong to the group of carotinoids. These are fat-soluble yellowred to red dyes that are stocked in certain cells (pigment cells/chromatophores) in the skin, under the epidermis (upper layer of the skin). *Crustaceae* have a higher constituent part of these dyes, which are stocked, to which - after having taken such life food - the fish then owe their strong red colouration.

About the following data of distribution of Dwarf Cichlids

It seems logic that data of distribution can only be given for those areas in which the spoken to Dwarf Cichlids have been found. But this does not have to mean that their range of distribution can reach much further than mentioned - this can at best be guessed, and given as a fact only when allocations (read: evidence) are available.

For example, in former decades it could happen that several members of species who are nowadays considered valid, were all interpreted as one species, treated in literature under this specific name and for them a wide distribution was given. So for example in his work from 1954 FOWLER - referring to statements of other scientists - for *A. taeniatum* (GÜNTHER, 1862) mentioned a distribution from Colonia Risso (Paraguay) and Cáceres (Mato Grosso) through the Rio Tapajós, Óbidos, Manáus and Tefé up to Tabatinga at the Colombia border. Today we know only a few *Apistogramma* species like *A. agassizii* which have an extended distribution, but they occur in several geographic varieties.

When in 1980 KULLANDER described a number of Dwarf Cichlids of the collection of the former king Léopold III. of Belgium and his companion, the Belgian Ichthyologist Dr J.-P. GOSSE, from the lower Rio Uaupés in northwest Brazil, he could only mention the Rio Uaupés as known distribution range, as type locality. Meanwhile, some specimens were also introduced from the system of the central to upper Rio Negro (*A. brevis*). It is similar to the specimens which presently are all considered *A. uaupesi*, like for example the two species ›Rotkeil-Apistogramma‹ and ›Segelflossen -Apistogramma‹. This listing is incomplete and in the future further collecting reports will have to be studied attentively.

Scientifically described and undescribed species
(SD = Sexual differences)

Apistogramma aequipinnis AHL, 1938
Meaning of the specific name: As much as ›with regular fins‹.
First description: Zool. Anz. 123, Berlin, pages 246-247.
Synonyms: None.
Distribution: Following the then still imprecise data ›supposedly Argentina‹.
Remark 1: The author (AHL) indicates: »From Mr W. Reitzig of Berlin, an interesting Dwarf Cichlid of the genus *Apistogramma* was sent to me which after the carried out determination turned out undescribed.«
Remark 2: In his taxonomic study of the genus *Apistogramma* with a revision of the Brazilian and Peruvian species, KULLANDER (page 15) includes a list of described *Apistogramma* species in chronological order of their description, mentions the locality given by the author and adds the presently valid name. Here KULLANDER adopts the distribution »supposedly Argentina« from AHL. In another listing (page 25), a list of valid (because examined in the type material) specific names with information on the distribution is included. Concerning ›Origin‹ it is written »Rio Paraguay (?)« (comp. text for *A.* spec. ›Paraguay III‹ as well as map on page 46).
Total length of the holotype 50 mm.

Apistogramma agassizii (STEINDACHNER, 1875)
Meaning of the specific name (etymology): In honour of J. L. R. Agassiz, born in Switzerland (Motier) in 1807, who studied in Heidelberg and after years of learning but also of disappointments emigrated to the USA where in 1848 he was called to Harvard University as professor. He was already 58 years old when he went off to his first trip to Brazil. In his preceding work at Munich, where together with J.B. de Spix he had to evaluate the fish collection of the botanic K.F.P. von Martius, the fish of the Amazon river were his first scientific subject. Later on, Agassiz collected a plenty of ichthyologically interesting material before he died in 1873.

S03322-4 *Apistogramma agassizii* ›Porto Velho‹, rivaling males.

S03310-3 *Apistogramma agassizii*, young wild male, blue variant.

First description: Sitzber. Kais. Ak. Wiss. Math. Naturwiss. Klasse 71 (1), pages 111-115.

Synonyms: *Geophagus* (*Mesops*) *a*.

Distribution: From the upper Amazon in Perú, Province Loreto, Ucayali Basin (Jenaro Herrera, Colonia Angamos [red colour form]), surroundings of Iquitos with Rio Nanay [Santa Clara]) along the Amazon River (Yavarí, triple country triangle with the borders to Colombia/Perú/Brasil [Puerto Nariño, Lago Matamata], Tabatinga), Rio Içá (reaching into Colombia where the river is called Rio Putumayo, cp. map on pages 10/11), Tefé and the Rio Japurá (Paraná do Amaná/Lago Amaná) which flows into the Amazon on the other side and further to the west, about to the surroundings of Santarém (Rio Tapajós). *A. agassizii* was also caught near the mouth of the Amazon. If moreover, it is true (Römer pers. comm.) that the type material of the species *A. parva* from the Rio Capim (see there) which is considered uncertain is a specimen of *A. agassizii*, this hypothesis would be confirmed.

Due to newer information, the distribution of *A. agassizii* reaches out passing Santarém to the east - roughly to the zone of the mouth of the Amazon near Belém. Between the mentioned colour varieties blue, yellow and red there are corresponding transitional forms.

S03330-4 *Apistogramma agassizii*, adult male, red-tail variant.

S03344-4 *Apistogramma agassizii*, young wild male.

<u>Remark:</u> In 1993, WINDISCH presents some of his specimens from a collecting trip to the Rio Cuieras and Barcelos at the central Rio Negro. They could be considered *A. agassizii*, but also *A. gephyra*. The author cautiously calls them *A.* cf. *agassizii* because a scientific evidence for *A. agassizii* from the central Rio Negro is still missing but is well present for *A. gephyra*. In this concern, the mentioned author refers to a publication by STAECK (1987) where it is indicated: »As I could find out by observations at five localities in the area of the lower Rio Negro which partially are quite distant from each other, in Brazil there are further colour races of *Apistogramma agassizii*. A first examination of the material collected there even makes it appear doubtful if the status of a valid species of *Apistogramma gephyra* KULLANDER, 1980 can be maintained, because the fish caught by me in the area of the type locality were without exception identified as *A. agassizii* by KULLANDER« (cp. map on page 41).

<u>Habitat:</u> The specimens were found in different water types (black, white and clear water). But there is always a very low total hardness (<1 °d) and a correspondingly low µS value. The pH value is mostly around the neutral value or even some tenths below.

<u>SD and colouration:</u> Corresponding to this wide distribution range, it is clear that various colour forms of the males developed. Besides this, out of these forms capable breeders developed some colour varieties.

In the wild forms of the males one differentiates:

The blue-white colour form: specimens of this type come from the greater surroundings of the city of Manaús which is situated at the mouth of the Rio Negro. In this case it is the type described by STEINDACHNER.

The yellow colour form: forms with yellow colour components are reported from different places. They are attributed to the surroundings of Santarém (SCHMETTKAMP, 1982), but also in the Rio Tefé drainage, yellowish specimens occur, as they were described by KOSLOWSKI (1994).

The red colour form: it can be assumed that for most aquarists the red variety is the most popular because, in the aquaristic trade already the bree-ding form ›Rotschwanz-Agassizii‹ (picture) which is a further development of this variety, gains high sales volumes. This breeding variety was derived from the wild form whose origin shall be in the upper Colombian-Peruvian part of the Amazon system. Specimens of this origin already show a slightly red flamed caudal fin as - more faintly - can also be found in some varieties of the yellow form.

<u>Total length:</u> Male 9 cm, female 5 cm.

<u>Maintenance in the aquarium and reproduction:</u> According to their wide distribution range along

Biotope: Lago Tefé,
the sample was taken at the beginning of the lake. An influence of the Rio Solimões (= central Amazon) is probably still traceable.
Date: 7.11.1985
Time: 9.25 o'clock
Water movement: Hardly moving (inlet)
Water depth: 1.00 to 2.50 meters
Water colour: Light tea brown

<u>Laboratory results:</u>

pH-value	7.3 (s. above)
Electrical conductivity (25°C), µS/cm	43
Acid capacity up to pH 8.2 (p-value), mmol/l	n.p.[1]
Acid capacity up to pH 4.3 (m-value), mmol/l	0.30
As carbonate hardness, °d	0.8
Total Alkalis (Ca^{2+}/Mg^{2+}), mmol/l	0.03
As total hardness, °d	0.2
Calcium (Ca^{2+}), mg/l	0.74
Magnesium (Mg^{2+}), mg/l	0.26
Total Iron ($Fe^{2+/3+}$), mg/l	0.05
Manganese (Mn^{2+}), mg/l	n.p.[2]
Ammonium (NH_4^+), mg/l	1.9
Nitrite (NO_2^-), mg/l	0.10
Nitrate (NO_3^-), mg/l	n.p.[3]
Chloride (Cl^-), mg/l	4
Sulphate (SO_4^{2-}), mg/l	1
Total Phosphate (PO_4^{3-}), mg/l	n.p.[4]
Hydrogen Carbonate (HCO_3^-), mg/l	18
Carbonate (CO_3^{2-}), mg/l	n.p.[5]
Oxidation Potassium Permanganate $KMnO_4$ consume, mg/l	13

[1] not provable, determination limit 0.05 mg/l
[2] not provable, determination limit 0.02 mg/l
[3] not provable, determination limit 1.00 mg/l
[4] not provable, determination limit 0.04 mg/l
[5] not provable, determination limit 2.00 mg/l

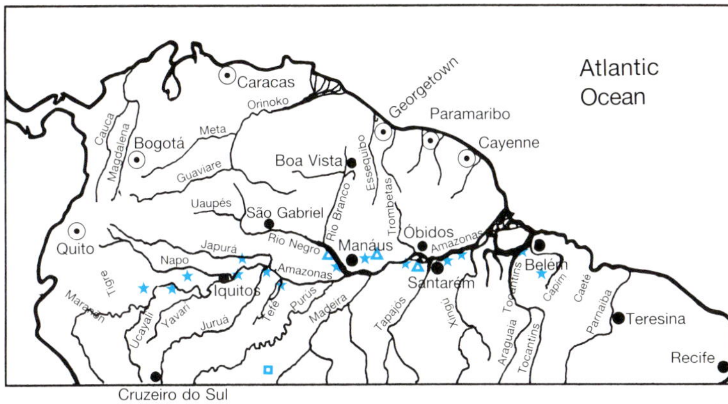

★ *A. agassizii*
□ *A. pulchra*
△ *A. gephyra*

the Amazon/Solimões, the specimens are best kept and reproduced in water that is as soft as possible, extremely low in nitrate and quite acidic (pH value below 6.0). As the males establish territories and can reach an impressive length, it is recommended to choose an aquarium with a length of more than 80 cm. If several females are kept, a sufficient number of hiding places, respectively breeding caves, has to be given. Sand is a recommendable ground, as it can be found on the soils of the Amazon area - indeed the upper layer shall consist of dark material (at least partly). Strong growth of aquatic plants in the rear part of the aquarium completes the general impression.

The representatives of this species also reproduce following the example of a male-mother-family in which the female alone takes care of the brood, whereas the partner provides security for the greater territory. As the female does not tolerate any cohabitant in the direct closeness of its breeding territory and ›opponents‹ are aggressively persecuted, in aquariums that are too small this can lead to unnecessary ›biting attacks‹ and therefore to an enormously enlarged expulsion of the male.

S03337-4 *Apistogramma agassizii*, young male, ›Santarém‹ black/red

The impressive delivery with far more than one hundred eggs develops in four days, to the hatch of the larvae and after five to six more days the young fish swim free. Now they are taken around for their first ingestion by the female now showing its most magnificent golden yellow colouration.

Remark: Unfortunately, today the type material which was deposited after a very short description cannot be found any more so that a control of the validity of the specific name is not given any more. These are the specimens whose late relatives were newly described as *A. moae* by KULLANDER, 1980.

Apistogramma amoena (COPE, 1872)

Meaning of the specific name: This is the adjective form of the latin word *amoenus* = ›lovely‹, ›beautiful‹. Following the grammar rules, the specific name was adapted to the female gender of the generic name.

First description: Proc. Acad. Nat. Sci. Philad. 1871 (1872), page 250.

Synonyms: *Geophagus a.*; *Geophagus taeniatus* (not GÜNTHER!) EIGENMANN & EIGENMANN, 1891; *Heterogramma taeniatum* EIGENMANN, 1919; *Apistogramma amoenus* REGAN, 1913.

Distribution: As distribution range the Rio Ampiyacú is given, a tributary of the Amazon in Perú coming from the north. REGAN (1913) describes a distribution in the drainage of the Rio Ucayali in Perú.

Apistogramma bitaeniata PELLEGRIN, 1936

Meaning of the specific name (etymology): From the latin words ›*bi*‹ = two and ›*taeniatus*‹ = with bands. The presently used ending ›a‹ has to be chosen because of ruling saying that the gender of the specific name in an adjective is to be adapted to that of the generic name.

First description: Bull. Soc. nation. Acclim. Prot. Nat. 83, pages 56-58.

Synonyms: *A. pertense* var. *bitaeniata*; *A. sweglesi* MEINKEN, 1961; *A. klausewitzi* MEINKEN, 1962; *A. kleei* MEINKEN.

Distribution: From the region around the well-known lake and river, the Rio/Lago Tefé (Igarapé do Ananas, Igarapé Preto) upstream the Amazon River (Solimões) passing the Colombian Leticia into the Peruvian province Loreto (Pu-

S03368-4 *Apistogramma bitaeniata*, wild male, yellow-brown variant.

S03362-3 *Apistogramma bitaeniata*, wild male, blue variant.

ñuisiqui Cocha, Zungaro Cocha, Tinaja, road Iquitos-Quistococha and others) as well as in the area of the city of Iquitos (Puerto Almendra) with the bed of the Rio Nanai (cp. map below).

Habitat: In most of the known habitats, these specimens occupy shallow zones of mixed and black water. The total hardness of the water can hardly be measured and therefore is below the 1°d-limit. Similarly low is the electric conductivity which in many habitats is between 10 and 20 µS. The pH value of the water is almost below the neutral value of 7.0.

SD and colouration: Contrary to the males, the females do not show an interesting finage. The caudal fin - in the partners bilobated, extended and with filaments - is only truncated.

A. bitaeniata occurs in the colour varieties yellow, blue and (less frequently) red. The specimens available through trade are mostly wild fish. The numerous importations of the recent years have shown that between each of the colour varieties mentioned above, there are even transitional forms.

Total length: Male 8 cm, female 5 cm.

Maintenance in the aquarium and reproduction: The specimens are kept similar to the *A. agassizii*. The beautiful fish do not belong to those that are easy to breed. The caudal fin of sexually mature females is truncated. They are only half as prolific as those of the previously mentioned species. Here we have a male-mother-family once more. For better development of the young fish a repeated water change per week is recommended, whereby the water quality should remain as constant as possible. The attractive finnage of the males develops the desired way only in very soft and acidic water with a low nitrate concentration.

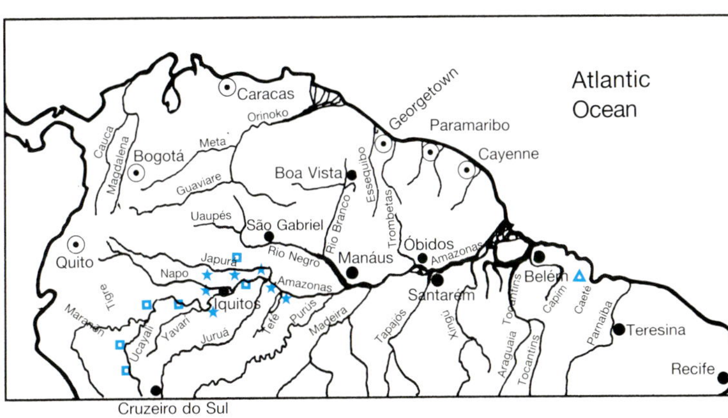

★ *A. bitaeniata*
□ *A. eunotus*
△ *A. caetei*

S03364-4 *Apistogramma bitaeniata*, wild male, red variant.

The relatively small females normally spawn in a cave. If this is not present, in the manner of other Cichlids that hide their eggs, they search for cracks in rocks where they can lay the eggs. Then the fertilizing males have no other choice but to set their sperm free in front of the hiding place and push it into the entry with the movements of their fins.

The future mother has now taken a lemon yellow covering with projecting black markings. Following the principle of male-mother-family, the female takes intense care of the brood whose larvae - depending on the water temperature - hatch four days after the fertilization. The transfer of the fry which in specimens of this genus can be observed frequently, is not done by all females with identical intensity. After some 5 to 6 days, the young fish swim free and one can now see that the intense colouration of the females produces a signal effect on her young. Danger is indicated by the mother with a jerky movement of the head, or also by quickly swimming fowards. The fry's reaction is similar to the one known from other Cichlids. Shocked, they sink to the ground and stay there motionless. When the danger is gone, the female swims above the bunch of fry, just as if she would wake up the dead.

Contrary to the behaviour of other pairs in which the male is later allowed to participate in the leading of the young, in this species, such a behaviour could not be observed. Each time the male came too close to the fry, it was chased off again and again. If females of the same species are in the breeding tank, there can be occasional jealousy among the specimens, and the females that remained without fry suddenly take over a yellow colouration in order to drive the leading mother away from her young by aimed attacks. But most of the time the mother is more aggressive and wins in the end.

Apistogramma borellii (Regan, 1906)

Meaning of the specific name (etymology): Dedicaction name in honour of the collector Dr A. Borelli. He discovered the first specimens for science and supplied the author with preserved specimens.

First description: Ann. Mag. Nat. Hist. (7) 17, pages 63-64.

Presently as synonyms are considered: *Heterogramma b.*; *H. ritense* Haseman, 1911 (?); *H. rondoni* de Miranda-Ribeiro, 1918; *A. reitzigi* Ahl, 1939.

Distribution: From the Brazilian Mato Grosso (Cuiabá at the upper Rio Paraguai [with Rio Cuiabá]) along the entire river up to where it flows into the Rio Paraná and in its dainage up to the north of Argentina. As the presently southernmost point of distribution flat water zones in the northern Argentinian province Entre Rios

Biotope: Rio Arara,
near the city of Leticia /south Colombia
Date: 4.3.1987
Time: 16.30 o'clock
Water movement: Slowly flowing
Water depth: Not established
Water colour: Light tea colour

Laboratory results:	
pH-value	5.90
Electrical conductivity (25°C), µS/cm	10
Acid capacity up to pH 8.2 (p-value), mmol/l	n.p.
Acid capacity up to pH 4.3 (m-value), mmol/l	0.07
As carbonate hardness, °d	0.20
Total Alkalis (Ca^{2+}/Mg^{2+}), mmol/l	0.03
As total hardness, °d	0.14
Sodium (Na^+), mg/l	1.8
Potassium (K^+), mg/l	0.65
Calcium (Ca^{2+}), mg/l	0.64
Magnesium (Mg^{2+}), mg/l	0.21
Total Iron ($Fe^{2+/3+}$), mg/l	0.58
Manganese (Mn^{2+}), mg/l	n.p.
Ammonium (NH_4^+), mg/l	0.03
Nitrite (NO_2^-), mg/l	n.p.
Nitrate (NO_3^-), mg/l	n.p.
Chloride (Cl^-), mg/l	1
Sulphate (SO_4^{2-}), mg/l	5
Total Phosphate (PO_4^{3-}), mg/l	n.p.
Hydrogen Carbonate (HCO_3^-), mg/l	4
Oxidation	
a) As Potassium Permanganate $KMnO_4$ consume, mg/l	6
b) As Oxygen (O_2), mg/l	1.5

n.p. = not provable

S03388-3 *Apistogramma borellii*, wild male, blue variant.

(= ›between the rivers‹ [Rio Paraná and Rio Uruguay] are known (LÜLING, 1977 and 1979) as well as ponds in the surroundings of the two cities Corrientes and Resistencia (connected by a big bridge across the Rio Paraná; MAYLAND, 1980). The swampy areas or ›Esteros‹ in the area between the two big white water rivers, form clear water lagoons in which *A. borellii* occurs and these are considered the main distribution range of this species. During the southern winter, (June to August) the author measured there the air temperatures between 16.0 and 17.3°C, but for example in the ›severe winter years‹ (1951 to 1960) temperatures between 2.9 and 0.1°C were possible, whereas summer temperatures between 26.2 and 27.3°C can be considered normal. Extreme maximum temperatures are given as 40.6 to 41.4°C. It is understandable that these air figures also influence the relatively shallow water (comp. map page 46).

Habitat: The habitat can be quite different. MAYLAND found these specimens in relatively shallow gullies whose water was covered especially by *Eichhorina azurea*. The capture had to be done by local fishermen who first of all submerged a big net deeply with a frame made of plumbing pipe (about 250 x 70 cm) and then had to take out the overflowing masses of the swimming plants in order to see the caught Dwarf Cichlids. The specimens found here had a deep blue basic colouration with a golden head and throat area.

Biotope: North Argentina,
drainage in the large arch of the Rio Paraná, Laguana Itati, east of the city of Corrientes, after Lüling (1977/79), carried out in the Hygienisches Institut at Bonn/Germany
Date: None
Time: None
Water transparency: Clear
Water colour: Colourless to slightly amber

Field measurements:	
Total hardness	1.3-1.4°d
Carbonate hardness	1.0°d
pH-value	6.8-7.1
Nitrite NO_2^-)	0.01-0.02 mg/l

Laboratory results:	
pH-value	6.8
Electrical conductivity µS/cm	38.9
Carbonate hardness, °d	0.8
As total hardness, °d	4.0
Calcium (Ca^{2+}), mg/l	1.2
Total Iron ($Fe^{2+/3+}$), mg/l	0.0
Manganese (Mn^{2+}), mg/l	<0.0
Nitrite (NO_2^-), mg/l	0.0
Nitrate (NO_3^-), mg/l	3.5
Ammonium (NH_4^+), mg/l	0.3
Chloride (Cl^-), mg/l	7.1
Sulphate (SO_4^{2-}), mg/l	1.0
Phosphate (PO_4^{3-}), mg/l	0.0
Oxidation Potassium Permanganate	
$KMnO_4$ consume, mg/l	18.2

S03389-4 *Apistogramma borellii*, wild specimens (males and females), blue-yellow variant, directly after capture.

LÜLING and companions caught the specimens with big hand nets and two man seines. As plants that can occasionally be found in the lagoons, *Cyperus californicus* and *C. reflexus* as well as *Limnanthenum humboldtianum* and likewise *Eichhornia azurea* are named.

In the specimens living further to the north (Rio Paraguai, Rio Cuiabá) the body colouration can be more or less intensely blue, instead the lower body part shows patterns of bright red stripes and spots (see ›Blue Form‹).

SD and colouration: Female specimens are plainly yellowish and show a by far more modest finnage. Compared to the colouration that, as described before, depends on the geographical origin, in the aquarium mainly two male colour forms are kept:

Specimens of the yellow form have a more or less golden yellow head and breast part as well as a deep blue body.

In the blue form, all the yellow parts are strongly reduced; instead a blood red head pattern is shown. Particularly well coloured males are also offered under the amendment "Opal". As *A. borellii* is considered a polychromatic* species, it can not be said with certainty if these are only selected specimens or pedigrees. Calling them a valid species for sure is not correct!

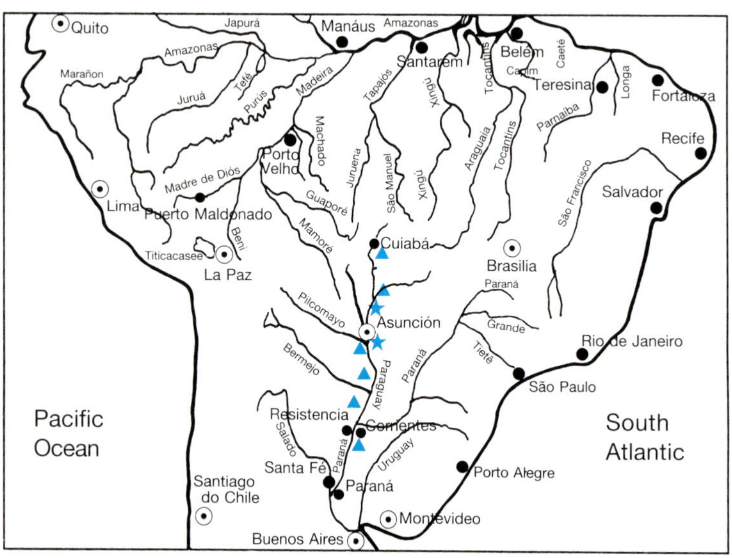

★ *A. spec. „Paraguay III"*
▲ *A. borelli*

In the hobby, yellow and blue strains are kept and developed further concerning colouration.

Total length: Male 6 to 7 cm, female maximum 4.5 cm. LÜLING (1979): »quite some 9 cm long«.

Maintenance in the aquarium and reproduction: In many habitats of this species, a rich aquatic vegetation can be found - in the water and at its surface. Therefore an aquarium also containing *A. borellii* should be furnished correspondingly. For these Cichlids that remain relatively small, an aquarium with about 60 cm length is sufficient. Bringing in cave-like hiding places made of roots and stones is also important for the well-being of these fish. Although several males in an aquarium constantly perform fights - mostly because of hierarchy - serious injuries are hardly inflicted. Even if the water in the habitats of the fish can still be called ›very soft‹, these Dwarf Cichlids prove to be adaptive and hence almost problem free. Similar facts can also be stated for the pH value of the water: it does not have to be in the low acidic range, but it can even be a bit above the neutral point of 7.0.

On the other hand, for successful reproduction it is recommended to take care of a low nitrate concentration and not to neglect regular water changes. In the southern part of their distribution range, the specimens of the species which even today by many aquarists is incorrectly kept under its synonym, *A. reitzigi*, have to tolerate seasonally conditioned cooler temperatures from 22 to 20°C. Indeed, for reproduction, these low temperature figures do not suffice so that the temperature has to be about 27°C. During the breeding period, the female is mostly not too aggressive towards the male and it so happens that the father is also endeavoured early with leading the fry, and this is allowed by the mother. After the common egg development period - four to five days up to the hatching of the larvae and five to six more days up to free swimming of the young fish - they immediately take *Artemia* nauplii and thereupon develop very well (do not forget continual water changes!).

Apistogramma brevis KULLANDER, 1980

Meaning of the specific name (etymology): In Latin, ›*brevis*‹ means ›short‹, so as much as ›small‹. In his description of the Uaupesi Dwarf Cichlids, KULLANDER selected this name because in this case he had fish of the smallest species.

First description: Bonner Zool. Monograph. (14), pages 107-111.

Synonyms: None.

Distribution: The type material comes from the collection of the former Belgian King Leopold III. who, together with the Belgian ichthyologist Dr J.-P. GOSSE toured the Uaupés area in the northwestern part of Brazil near the border to Colombia and there he found the specimens in the area of the Rio Uaupés (Rio Vaupés in Colombia), more exactly in the Rio Tiquié (an

S03397-3 *Apistogramma* cf. *brevis*, male.

S03397-3
Apistogramma cf. *brevis*, female.

affluent coming from the south west), in the Igarapé Acaraposo, in the Rio Uaupés itself (near the towns of Trovão and Assaí) as well as in the Lago Penera. The holotype (a male) had a SL of only 29 mm (which means measured without the caudal fin) and comes from a small Igarapé (= gully or stream) to the Lago Penera. These small Dwarf Cichlids - as is already expressed by the specific name - do not come to us with common importation. At best, some specimens are introduced in small numbers by active interested aquarists themselves. But in fact it has to be assumed that regularly some specimens - but in most cases unrecognized - are spread all over the world as a supplementary catch. Due to our observations, these supplementary catches are limited to those import shipments reaching us, together with *Dicrossus filamentosus* via Manáus. In these shipments, there are often young fish of *A. pertensis*, *A. uaupesi* and *A. brevis*. As at this stage the specimens are still very small, *A. brevis* are often confused with *A. pertensis*. But a more detailed study shows that these specimens attract attention because of their more compressed and deeper body shape (cp. map on page 61).
Habitat: During most times of the year when there is a high water level, Dwarf Cichlids are especially found in inundated zones where submersely vegetating grass-like terrestric plants offer shelter to the specimens. Here the water is extremely poor in minerals (conductivity about 10 µS) and with a pH value around 4.8, it is strongly acidic.
SD and colouration: With a body length of only 2.5 cm the optical separation of the sexes starts already. In the males, the at first rounded caudal fin starts to become frayed and flatened vertically. The transition from the rounded part of the caudal into the straight vertical part slowly develops extensions. The females keep the round form of the caudal fin. In state of excitement, the relatively dark lateral band can be replaced by five likewise strong vertical bands. The occasionally visible round lateral blotch is slightly bigger in diameter than the width of the lateral band. The external appearance of this small but rare species is limited so that one can assume that only a few specialized amateurs want to keep and reproduce these dwarfs.
Total length: Following KULLANDER, the biggest of the collected specimen (a male) has 38.7/55.4 mm, the biggest female has 27.1/36 mm.
Maintenance in the aquarium and reproduction: Cp. *A. diplotaenia*.

Apistogramma cacatuoides HOEDEMAN, 1951
Meaning of the specific name (etymology): From ›*cacatu*‹ = cockatoo and ›*oides*‹ (Greek) = similar. The anterior dorsal fin rays which are errected in the males remind one of the crest of a cockatoo.

First description: Beaufortia 1 (4), pages 1-4.

Remark: In the first description, the origin of the holotype was erroneously given as ›near Paramaribo‹ (the capital of Surinam) in the north west of South America. As remarked by KULLANDER (1989, page 83), the error must have been caused when two males of *A. steindachneri* were in the original description with the same data about locality and registration number having the name *A. cacatuoides* (ZMA 100.033). Among two of the three labels in the bottle with *A. steindachneri*, both written with ink, one indicated the locality ›Dutch-Guiana (ditch near Paramaribo)‹ and had the manuscript name ›Hoedeman 1950‹). The third label, written with pencil, was wearing the name ›*Apistogramma cacatuoides* Hoedeman 1951‹. The corresponding ZMA numbers could be read badly by which the confusion can probably be explained.

Synonyms: Due to a wrong description the species was occasionally called *A. borellii*.

Distribution: In the Peruvian part of the Rio Ucayali and the Amazon. In the Brazilian part almost up to the border triangle Perú/Colombia/ Brazil near Tabatinga.

Proven localities in Perú are the waters in the area of the rivers Urubamba and Tambo, which unite near the city of Chicosa and form the Rio Ucayali. Furthermore at the Rio Huánuco along the road to Puerto Inca as well as along the road Pucallpa-Tingo María and also in the area of the Rio Pachitea. In the drainage of the Rio Ucayali in the surroundings of the Yarina Cocha, a pond at the road to Campo Verde-Tournavista and others.

In the drainage of the Rio Samiria in a gully near Chinguito Cocha. In the Amazon drainage in the surroundings of the city of Iquitos as well as at the Rio Ampiyacu. In the bed of the Rio Yavarí in the Lago Matamata as well as in the Caño Piraña. In the Colombian Amazon in the surroundings of the city of Leticia (Isla Santa Sofia and Puerto Nariño, Lago Tarapoto).

In recent times, very beautiful specimens from an area near the Peruvian/Colombian border, in fact from the drainage of the Rio Putumayo, are offered (cp. map on page 51).

Habitat: The waters in the areas close to the Ucayali transport minerals (not only calcium oxide corrosives!) from the Andine regions and

S03406-4 *Apistogramma cacatuoides*, orange-blue pedigree.

S03404-4 *Apistogramma cacatuoides*, pedigree ›Doppelrot‹ .

therefore they show slightly different values. For example the total hardness at the upper Rio Ucayali has 10 to 12° and in some areas it is even surpassed by the carbonate hardness for some degrees. The conductivity is at about 400 µS and the pH value moved into the alcaline range from 7.4 to 7.6. Indeed, here the nitrate value which in the Amazon area is commonly low can neither be measured by <1. These water values let us see the known adaptability of *A. cacatuoides*, therewith also the most willing to reproduce and the resulting various colourful pedigrees which mainly means an improvement of the colour appearance.

SD and colouration: Contrary to their male partners, female specimens show only a normal dorsal fin as well as a truncated caudal fin. Males with elongated membranes of the anterior dorsal fin rays. Ends of the dorsal, anal and ventral fins are elongated. In adult specimens, the caudal fin is bilobated. In the more colourful male offspring the long fin extensions are partly degenerated.

In the various areas of the vast distribution range one can find males of moderate, green-gray colouration, but also with red dots or orange zones in the fins. Already for a long time from the eastern part of Germany as well as from other countries offspring males of colour forms like ›gold‹, ›yellow‹, ›blue‹, ›green‹, ›red‹ as well as ›double red‹ and ›triple red‹ are offered.

Total length: Male 8 cm, female 5 cm.

Maintenance in the aquarium and reproduction: Compared to Dwarf Cichlids living in softer water, considerable differences in maintenance and breeding only exist in the fact that the specimens like *A. cacatuoides* with their origin in waters of certain Andine slopes, can adapt to

Biotope: Lago Tarapoto,
south Colombia
Date: 13.3.1987
Time: 14.00 o'clock
Water movement: Hardly measurable
Water depth: Not established
Water colour: Tea coloured (light black water)

Laboratory results:

pH-value	6.72
Electrical conductivity (25°C), µS/cm	116
Acid capacity up to pH 8.2 (p-value), mmol/l	n.p.
Acid capacity up to pH 4.3 (m-value), mmol/l	0.99
As carbonate hardness, °d	2.78
Total Alkalis (Ca^{2+}/Mg^{2+}), mmol/l	0.51
As total hardness, °d	2.84
Sodium (Na^{+}), mg/l	3.4
Potassium (K^{+}), mg/l	1.6
Calcium (Ca^{2+}), mg/l	17
Magnesium (Mg^{2+}), mg/l	2.0
Total Iron ($Fe^{2+/3+}$), mg/l	0.21
Manganese (Mn^{2+}), mg/l	n.p.
Ammonium (NH_4^{+}), mg/l	0.08
Nitrite (NO_2^{-}), mg/l	n.p.
Nitrate (NO_3^{-}), mg/l	n.p.
Chloride (Cl^{-}), mg/l	3
Sulphate (SO_4^{2-}), mg/l	6
Total Phosphate (PO_4^{3-}), mg/l	0.04
Hydrogen Carbonate (HCO_3^{-}), mg/l	60
Oxidation	
a) As Potassium Permanganate $KMnO_4$ consume, mg/l	16
b) As Oxygen (O_2), mg/l	3.0

n.p. = not provable

S03419-4 *Apistogramma cacatuoides*, wild form with solid red pattern in the caudal fin.

slightly higher hardness and pH values (whereby the latter can be some tenths above the neutral value). The specimens of this species, whose habitats are situated from the lower Rio Urubamba (Perú), downstream the Rio Ucayali, but also beyond its flowing together with the Rio Marañon and further downstream to the Amazon, show this adaptability.

With the help of the example of the Cockatoo Dwarf Cichlid, the reproduction can be explained well because it can also be initiated successfully by less specialized aquarists. For breeding this species, the water can have a total hardness up to about 12° and a pH value up to the neutral limit 7.0. The water temperature should be between 26 and 27°C. A regular, partial water change, done roughly every ten days, additionally contributes to the well-being of the specimens. Besides *A. agassizii* and *A. borellii*, also in *A. cacatuoides* particular colour forms like ›red‹, ›double red‹ and ›triple red‹, ›yellow‹ and ›orange‹ are created through reproductive selection. For reproduction, most aquarists appreciate the mentioned males with many correspondingly colourful blotches in the fins. Indeed, in some strains this leads to negative developments. For example, degenerative appearances and the thereby related omission of long fin extensions can be seen occasionally.

In the furnishings of the roughly 80 cm long aquarium containing various caves the hiding places should be far enough from each other (at least 25 cm) so that even several females can be kept. This is convenient for males of various species tending to the formation of harems. As ›caves‹ also half coconut shells (with a whole

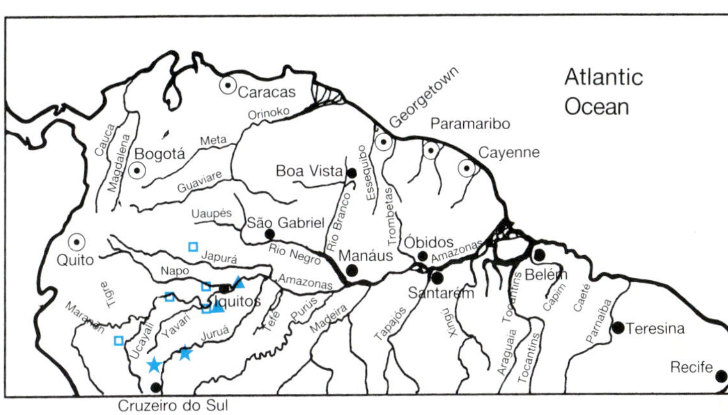

★ *A. juruensis*
□ *A. cacatuoides*
▲ *A. norberti*

to enter) can be added. But only one male can be in the tank because otherwise the weaker specimen would be often attacked or even killed by the superior one.

Females lay their eggs at the ceiling of the cave occupied by them and thereupon, take over alone the care for the brood. During this time, the male controls the territory in a particularly careful and aggressive way. If several females are in the aquarium, the polygamist has a lot of work to do and then seems to be particularly nervous. The larvae hatch after about 60 hours after fertilization and then they are stored in pits. Depending on the temperature and the coinciding larval development, the young fish swim free after 8 to 9 days. Their yolk sack is consumed and the independent ingestion starts, whereby the mother takes care of its offspring and leads them to the feeding place. The little mouths of the small fish are already big enough to take *Artemia* nauplii. In order to offer more variability to the young fish, one can offer a selfmade mash containg flake food. It simply consists of very fine food which is dissolved in aquarium water and given to the fry with the help of a pipette in order to be able to squeeze it well-directed towards the swarm. The care of the females will take on two more weeks, then the young dwarfs have to cope their own life.

Apistogramma caetei KULLANDER, 1980

Meaning of the specific name (etymology): According to the main river in the distribution range, the Rio Caeté.

First description: Bonner Zool. Mongraph. (14), pages 76 to 79.

Synonyms: None.

Distribution: As type locality, an Igarapé (river, gully) in Bragança, a town near the coast east of the city of Belém in the Brazilian federal state Pará, is given. Further description material comes from the small Rio Apeú near the town of Castanhal and further in the area around the Baia de Marajó (comp. map on page 43).

Habitat: In ditches, ponds and similar waters where foliage, wood and overhanging vegetation at the shores gives cover and therefore shelter to the specimens. The water is mostly very soft, but in this area it can lose its formerly good quality because of agricultural pollution.

S03430-3 *Apistogramma caetai*, wild male.

Location: habitat of *A. caetai* in the upper Rio Apeú.

SD and colouration: In semi-adult specimens, the sexes can not be differed well. After the beginning of sexual maturity the females adapt a yellow colouration which is intensified during the mating season and reproduction.

A species whose representatives - at least concerning the head patterns - show certain similarities with *A. commbrae*. It is this bloodred blotched pattern which adds a certain beauty to these specimens. Besides, an almost constantly shown black lateral band and the oval vertically oriented blotch on the base of the caudal fin are conspicuous. Contrary to that, forehead and snout stripes as well as the band on the cheeks only occur depending on the mood. Basic body colour is greenish yellow with blue glitter in the area of the dorsal, anal and caudal fins. The caudal of the males is slightly rounded and shows a rusty red dotted pattern arranged in an arched way.

Remark: The formation of the pedigree ›Bluehead-Apistogramma‹ started in South East Asia, but there it created almost exclusively males. It was the German breeder Werner KRIEG of Hildesheim who was able to build up a new breeding line with the help of a few Asian males and some wild females.

Total length: Male 6 cm, female 3.5 to 4.5 cm.

Biotope: Rio Apeú,
in a stream-like narrow water in the upper part of the river, east of Belém, Pará (Brazil), still however before Castanha
Date: 23.11.1996
Time: 11.15 o'clock
Water movement: Rapidly flowing
Water depth: Maximum 50 cm
Water colour: Clear and greenish with vegetation
Surface temperature: 28.9°C

Laboratory results:

pH-value	5.97
Electrical conductivity (25°C), µS/cm	18
Acid capacity up to pH 8.2 (p-value), mmol/l	n.p.
Acid capacity up to pH 4.3 (m-value), mmol/l	0.07
As carbonate hardness, °d	0.20
Total Alkalis, mmol/l	0.028
As total hardness, °d	0.16
Sodium (Na^+), mg/l	1.5
Potassium (K^+), mg/l	0.6
Calcium (Ca^{2+}), mg/l	0.59
Magnesium (Mg^{2+}), mg/l	0.33
Total Iron ($Fe^{2+/3+}$), mg/l	0.37
Manganese (Mn^{2+}), mg/l	0.005
Ammonium (NH_4^+), mg/l	<0.02
Nitrite (NO_2^-), mg/l	<0.02
Nitrate (NO_3^-), mg/l	0.6
Chloride (Cl^-), mg/l	2.3
Sulphate (SO_4^{2-}), mg/l	1.2
Phosphate (PO_4^{3-}), mg/l	<0.04
Hydrogen Carbonate (HCO_3^-), mg/l	6.1
Oxidation	
a) As $KMnO_4$ consume, mg/l	5.4
b) As Oxygen (O_2), mg/l	1.4

n.p. = not provable

S03436-4 *Apistogramma caetai*, pedigree ›Blaukopf‹.

Maintenance in the aquarium and reproduction: Specimens of this species are kept and bred in aquaria with a length between 60 and 80 cm which as usual are decorated with hiding places consisting of roots and stones. A mixture of fine gravel and sand should be used as a bottom. A rich background vegetation gives additional safety to the relatively small specimens.

When it comes to water quality, no exceptional requirements need to be fulfiled, but it should be as soft, and as poor in nitrate as possible, additionally it should be slightly acidic. Because of their natural habitat lying directly under the equator, a water temperature of 27°C is appropriate. The polygamous males form a harem with several females and mate with these, following the principle of the male-mother-family one after the other. The 50 to 60 eggs are deposited with their short stems of one pole on the ceiling of the cave. The development, to the hatching of the larvae takes about three days, after five more days the young fish start swimming free. The young are immediately able to eat larvae of *Artemia salina*.

Apistogramma commbrae (EIGENMANN in REGAN, 1906)

Meaning of the specific name (Etymology): The name is based upon a reading error of the author in a hand written letter by EIGENMANN (for ›Corumbá‹, a city by the Rio Paraguay). But the wrong name is valid as ›original spelling‹. A later correction (›Emendation‹) was not accepted.

First description: Ann. Mag. Nat. Hist. (7) 17, pp. 64-65.

Remarks: In his description, REGAN assertively refers to the cooperation with Prof EIGENMANN: »...which should be called *H. commbrae*, EIGENMANN.« The type material originates from the collection of Dr Borelli.

Synonyms: *Heterogramma corumbae* EIGENMANN, MCATEE & WARD (1907), p. 146; *Apistogramma corumbae* REGAN (1913), Ann. Mag. Nat. Hist. (8) 12, p. 282.

Distribution: The species can be found in the area of the Brazilian Mato Grosso from Cuiabá and Cáceres and further to the south up to the neighbouring federal state of Mato Grosso do Sul with the Pantanal abounding in water, and to the city of Corumbá along the Rio Paraguay up to the drainage of the Rio Paraná near the cities of Corrientes and Resistencia (northern Argentina), cp. map page 77.

Habitat: The small Cichlids prefer habitats rich in aquatic vegetation which in the areas mentioned are additionally shadowed by a dense layer of swimming plants. The water readings found by MAYLAND at various locations cannot be called extremely soft (conductivity about 100 µS) and the pH value is mostly a few tenths above neutral (7.0).

Biotope: Rio Cuiabá,
upstream the city of Cuiabá
Date: 7.11.1996
Time: 17.00 o'clock
Water movement: Sluggishly flowing
Water colour: carrying lots of sediments, cloudy
Surface temperature: 26.3°C

Laboratory results:

		Magnesium (Mg^{2+}), mg/l	2.9
		Total Iron ($Fe^{2+/3+}$), mg/l	0.17
pH-value	7.03	Manganese (Mn^{2+}), mg/l	0.002
Electrical conductivity (25°C), µS/cm	78	Ammonium (NH_4^+), mg/l	<0.02
Acid capacity up to pH 8.2 (p-value), mmol/l	n.p.	Nitrite (NO_2^-), mg/l	<0.02
		Nitrate (NO_3^-), mg/l	1.9
Acid capacity up to pH 4.3 (m-value), mmol/l	0.65	Chloride (Cl^-), mg/l	1.0
		Sulphate (SO_4^{2-}), mg/l	4.1
As carbonate hardness, °d	1.8	Phosphate (PO_4^{3-}), mg/l	0.08
Total Alkalis (Ca^{2+}/Mg^{2+}), mmol/l	0.33	Hydrogen Carbonate (HCO_3^-), mg/l	40
As total hardness, °d	0.78	Oxidation	
Sodium (Na^+), mg/l	1.7	a) As KMnO4 consume, mg/l	6.8
Potassium (K^+), mg/l	1.6	b) As Oxygen (O_2), mg/l	1.7
Calcium (Ca^{2+}), mg/l	8.3	n.p. = not provable	

S03445-3 *Apistogramma commbrae*, wild male.

SD and colouration: Specimens of this species are occasionally introduced with representatives of *A. borellii* and/or *A. trifasciata* (which are commonly considered more beautiful). Well kept fully adult males show a yellowish glimmer on a grey brown basic colouration. On the lateral head (cheeks and gill covers) patterns of blue and red highlights are conspicuous, just as these are known in a similar way from *A. caetei*. Red tones can be detected in the dorsal, caudal and anal fin.

Depending on the mood of the specimen, besides stripes on the forehead, the snout and the cheeks, vertical and horizontal bars are shown also. A lateral blotch as well as a caudal blotch have to be added. The blotch on the base of the caudal sometimes seems like some kind of a double blotch with the form of a lying, slightly compressed eight also two to three stripes on the lower belly can be seen.

During the spawning period, the females, which are normally yellow brown, develop a bright yolk yellow colouration which can occasionally be a blueish green, in which the breast and the belly display themselves in particularly intense tones. The fins are also a very intense yellow and the anterior rays of the dorsal and the pectorals become deep black. The lateral band is reduced to a narrow often discontinued line.

Total length: Males normally reach a total length of 5 cm, females can reach about 3 cm.

Maintenance in the aquarium and reproduction: One male can be socialized with two females, if the size of the breeding tank permits. If the male mates with both partners, there can be territorial fights between the females taking care of the brood, so that in such a case it is recommended to install the breeding caves as far away from each other as possible. Also, fighting females turn out not to be demure towards the male - the specimen frequently attacked from both sides should then be taken out of the tank.

Once the young fish are independent enough to escape from their mother, it can happen that without any problem they are protected by the other female. But the bigger the young grow, the more likely they are recognised as intruders into the swarm by the foreign female and thus ›separated‹ by nudges.

Concerning general maintenance, *A. commbrae* can be called a problem free Dwarf Cichlid which places relatively few demands on its keeper. Reproduction was successful in water of a total hardness of 8°d, a pH value around 7.0 as well as an average temperature of 24°C.

Apistogramma cruzi KULLANDER, 1986

Meaning of the specific name (etymology): A dedication name to the honour of the cooperator (›co-collector‹) and ›tireless helper‹ during collecting trips in 1981 (Pebas) and 1984 (Mazán): José Cruz Rodriguez.

S03457-3 *Apistogramma cruzi*, wild male.

First description: Cichlid fishes of the Amazon River drainage of Peru. Swed. Mus. Nat. Hist., Stockholm (431 pp.), pp. 159-163.

Synonyms: None.

Distribution: This very variable species can be found in the widely limited drainage of the Rio Napo in Ecuador and Perú (lower Rio Mazán upstream of Puerto Alegre and upper Rio Aguarico/Perú and San Pueblo de Kantesiya at the Rio Caimito/Ecuador) as well as in the area of the Rio Putumayo drainage in Colombia (Rio Caucayá [Limon Cocha] and of the Rio Caquetá and Rio Caguán drainge), cp. map page 56.

Habitat: With the example of the Rio Mazán, the habitat of a drying up stream with a width of about 3 m and a depth of 50 cm could be depicted briefly by the author. Its water was quite colourless, the muddy bottom was littered with foliage, branches and twigs.

SD and colouration: During the breeding period, female specimens show a golden yellow colouration. On the olive grey basic tone of the males, there is a pattern of hinted vertical bars, a mostly quadratic lateral band and two to three stripes on the lower belly. On the head pattern with forehead, snout and cheek stripes, in the

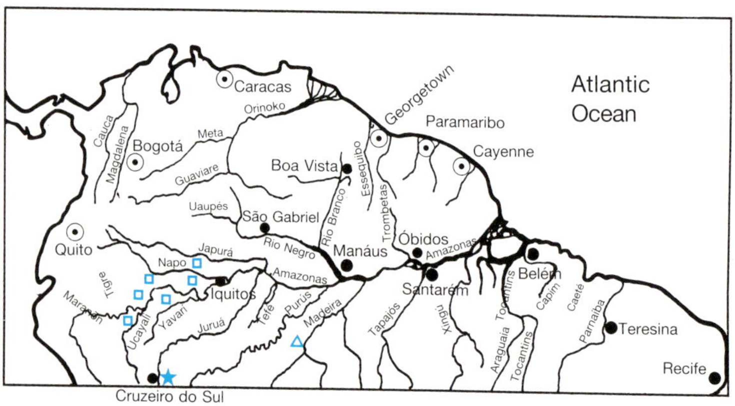

★ *A. moae*
□ *A. cruzi*
△ *A. resticulosa*

S03459-4 *Apistogramma cruzi*, wild male, blue variant.

lower part a weak pattern of several sky blue lines can be seen. Weak blue tones can also be seen in the dorsal and anal fin which (just like the caudal fin) also show the basic colouration mentioned before.

New Import in 1996.

Total length: Male 8 cm, female 5.5 cm.

Maintenance in the aquarium and reproduction: In the course of new imports from Perú in June 1996, specimens with a particularly beautiful pastel colouration in the colours of the rainbow were introduced. According to their origin, in maintenance and breeding it belongs to the less difficult species and is similar to *A. cacatuoides*.

S03459-4 *Apistogramma cruzi*, wild female.

S03455-4 *Apistogramma cruzi*, wild male, beige variant.

Apistogramma diplotaenia KULLANDER, 1987

Meaning of the specific name (etymology): From ›*diplos*‹ = double and ›*taenia*‹ = band, stripe: with a double band. Before their scientific description, specimens of this species were designated with the trivial name ›Doppelband-Apistogramma‹.

First description: Zool. Scripta 16 (3), pp. 259-270.

Synonyms: None.

Distribution: As far as could be established, these small Cichlids are found at many places in the shallow water zones of the drainage of the central Rio Negro, but there they only rarely occur in larger numbers.

Collections in the area around the confluence of the Rio Padauari with the Rio Negro (61.1°W) north of the mouth of the Rio Negro (Rio Arirahá) have been proven, where the fish occur together with a small *Crenicichla* species (probably *C. regani*), *Apistogramma regani*, *A. gibbiceps* and *A. paucisquamis* as well as with a *Dicrossus* species (›Doppelpunkt‹- or ›Rio-Negro‹). Until now, this species is still scientifically undescribed.

On the other hand, the localities given by KULLANDER (1987) are further to the northwest in the upper Rio Negro, partly on Venezualian territory (San Carlos do Rio Negro in the southwestern corner of the country) as well as in the area of Urubaxí and Daraá at the mouth of the Rio Arirará. One of the authors (LINKE & STAECK, 1995) »could catch this Cichlid in the area of the Anavilhanas Island Group where it is indeed rare«. In the latter case, the accompanying fauna consists of *A. pertensis*, *A. gephyra*, *Acarichthys heckelii* as well as *Mesonauta insignis*.

The water of the Rio Negro is not only deep brown respectively almost black; despite its darkness it is extremely clear! The total hardness can not be detected with the test kits and devices used in the hobby and for the water sample given here, it was professionally tested and the value was given as 0.003. The pH value is also remarkably low and the water is even more acidic in the lower part of the river (cp. water sample ›Anavilhanas Island Group‹, page 100 and map page 59).

Habitat: In waters like those described because of the black colouration of the water there is no

S03465-4 *Apistogramma diplotaenia*, wild male, blue.

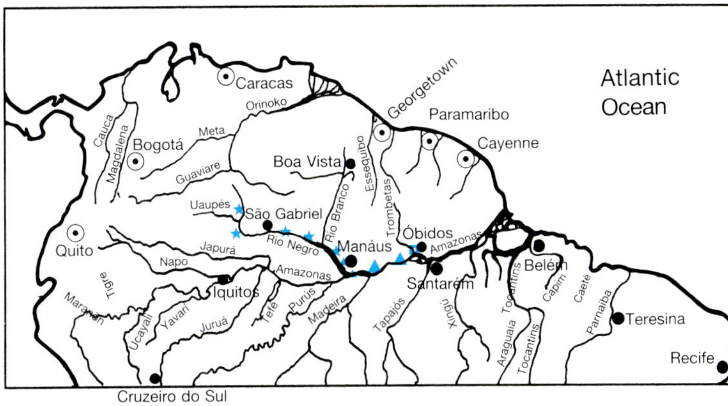

★ *A. diplotaenia*
□ *A. geisleri*
▲ *A. regani*

aquatic vegetation - at best near the banks in the shallow water there is overhanging vegetation. The bottom consists of a fine bright sand, but it is often covered by a thick layer of foliage and branches where the fish build their hiding places because the well known predators chasing the small Cichlids (in this concern these are mainly the semi-adult Characins of the genus *Hoplias*) are here also present. The water samples taken in the area near the confluence of the Rio Preto and the Rio Padauari which was mentioned before (WINDISCH, 1991), showed a conductivity of less than 10 µS and a pH value of 4.2. At the locality of *A. diplotaenia*, the water was about 1.60 m deep.

SD and colouration: If kept in pairs in very small aquaria, the males of the Dwarf Cichlids first presented under the trivial name ›Doppelband-Apistogramma‹ by SCHMETTKAMP in 1981, can become extremely aggressive. Therefore, it is advisable to keep *A. diplotaenia* in pairs without further cohabitants: The average aquarist will most likely never possess this species, the specialist will proceed as mentioned. In the state of affection, their normally sand yellow colouration gives way to an almost white colouration above which then - particularly in the dorsal, anal and pectoral fins - a more or less intense blue glimmer develops (photo) which at the border of the dorsal fin condenses into a marginal (= outer) band. The two strong horizontal bands cannot be ignored and therefore helped the species with its name. The relatively big eyes let us perceive an adaption to the dark habitat. Female specimens are often smaller and in their colouration they appear more modest, in the natural habitat one mostly finds specimens of both sexes in which the double horizontal band is replaced by a single dark band. Due to the relatively wide distribution range in the drainage of the Rio Negro from the Anavilhanas (about 80 km north of Manáus) up to the Venezualian border (San Carlos), it is not astonishing that several geographical colour varieties developed of which the one displayed herein is considered as one of the most beautiful ones.

Total length: Male 5 cm, female 3.5 to 4 cm .

Biotope: Upper Rio Negro,
behind the rapids at a sandbank upstream the city of São Gabriel da Cachoeira (formerly Uaupés)
Date: 17.10.1986
Time: 6.45 o'clock
Water movement: Rapidly flowing
Water depth: 150 cm - decreasing
Water colour: Darkest of all
Water sample: Strong tea brown

Field measurements:	
pH-value	6.220
Oxygen (O_2)	7.24
Water temperature	27.1°C
Air temperature	26.0°C
Laboratory results:	
pH-value	4.42
Electrical conductivity (25°C), µS/cm	15
Acid capacity up to pH 8.2 (p-value), mmol/l	<0.05
Acid capacity up to pH 4.3 (m-value), mmol/l	0.02
As carbonate hardness, °d	0.1
Total Alkalis (Ca^{2+}/Mg^{2+}), mmol/l	0.06
As total hardness, °d	0.003
Calcium (Ca^{2+}), mg/l	0.11
Magnesium (Mg^{2+}), mg/l	0.056
Total Iron ($Fe^{2+/3+}$), mg/l	0.16
Manganese (Mn^{2+}), mg/l	<0.02
Ammonium (NH_4^+), mg/l	0.08
Nitrite (NO_2^-), mg/l	<0.02
Nitrate (NO_3^-), mg/l	<1
Chloride (Cl^-), mg/l	<3
Sulphate (SO_4^{2-}), mg/l	7
Total Phosphate (PO_4^{3-}), mg/l	0.06
Hydrogen Carbonate (HCO_3^-), mg/l	1
Oxidation Potassium Permanganate $KMnO_4$ consume, mg/l	321

Maintenance in the aquarium and reproduction: The small specimens which are particularly impressive due to their double horizontal band, puzzled many aquarists for longtime maintenance and breeding attempts. For various breeders it turned out that it is advantageous to keep the specimens in bigger groups (if these are available) in correspondingly wider aquaria with up to 140 cm length. Immediately after being placed into the tank, the males begin to orientate themselves and occupy territories that are interesting for them. Disputes occur, but these - favoured by sufficient space - mostly do not end with the death of an inferior specimen. The occasional threatening of the territory border is a part of the common ritual.

For reproduction, the lowest total hardness and a quite low pH value are indispensable. They can correspond to the data given for the water quality and also the installation in *A. elizabethae*. Now the specimens have to be fed well and in a diversified way, thereby *Crustaceae* are preferred. Whoever observes the females more frequently, will soon discover the presence of eggs. While mating, the males are not particularly persevering. A female ready to spawn will soon follow its mating partner to the entry of the cave. If on the other side it does not follow, it is driven away without further ado. First of all, the male is swimming into the selected cave as if it wanted to inspect it thoroughly. If it returns, the female occupies the cave and soon starts depositing the whitish to sand coloured eggs. If a packet of foliage was selected for a spawning cave, also here the eggs are fastened to the ceiling (made of leaves). At 27°C, the larvae develop in the known three days until hatching and after four more days, the young swim free.

Apistogramma elizabethae KULLANDER, 1980

Meaning of the specific name (etymology): Dedication name. Ms Elizabeth Cabot Cary Agassiz (1822-1907), in whose honour the species was named, was the second wife of Prof Louis R. Agassiz, participant of the Thayer expedition (1865-1866) and main author of the book about this trip (AGASSIZ & AGASSIZ, 1868). Corresponding to a remark of the author KULLANDER, the name shall reflect the similarity of the new species to *A. agassizii* from the Amazon.

S03495-4 *Apistogramma elizabethae*, male, yellow variant.

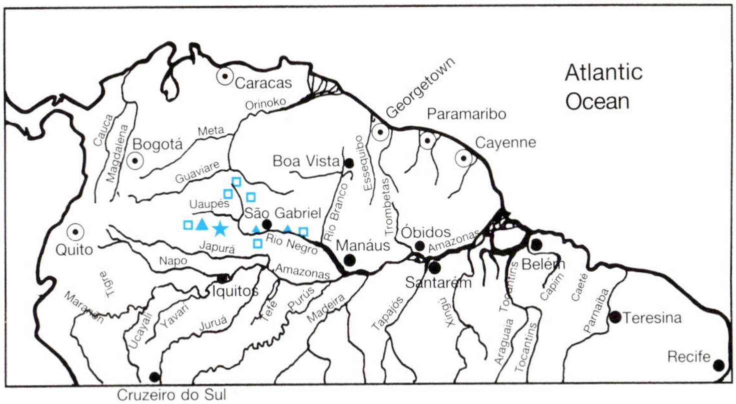

★ *A. elizabethae*
□ *A. uaupesi*
▲ *A. brevis*

First description: Bonner Zool. Monograph. (14), pages 103-106.

Synonyms: None.

Distribution: Hitherto known only insufficiently. The localities known since today originate from the area of the Rio Uaupés (in Colombia Rio Vaupés). The holotype comes from an Igarapé, a right sided affluent near the town of Travão (about 20 km away from the mouth of the river into the Rio Negro, 00°02'N, 67°26'W). Further occurences were proved in the neighbouring Lago Penera where the fish lives sympatrically (together in the same habitat) with *A. brevis*, *A. meinkeni* and *A. uaupesi* (comp. map above).

Habitat: These Cichlids were met in zones of slowly flowing water near the banks with deposits of foliage in very acidic water (pH 4.2 to 5.2). The water temperature at the surface at noon was approximately between 26 and 29°C. The average figures of the total hardness measured were about 10°.

SD and colouration: In DCG-Info 24 (8), 1993, Römer *et al.* present these fish with a picture and report about two colour types, a yellow grey and a red blue form. *Apistogramma elizabethae* is a polychromatic species.

The relatively slender males show a wide horizontal band which depends on their mood, running from the posterior edge of the eye all through the caudal peduncle into the lancet like caudal fin which is almost colourless. In their central part, the yellowish ventral fins have a deep black pattern. If the mentioned horizontal band remains pale, the two deep black round lateral blotches behind the beginning of the pectoral fins can be seen particularly well. Slanting stripes on the cheeks are present. Breeding females show deep black ventral fins.

The extended membranes of the third ray of the dorsal fin in the male has to be considered particularly remarkable.

Total length: The biggest specimen among the type material, a male, had a standard length (SL = measured without caudal fin) of 39.8 mm (which would correspond to a total length of about 57 mm), but the biggest female was only 23.8 mm (SL) long (TL = ±32 mm).

Maintenance in the aquarium and reproduction: For the inhabitants of relatively acidic and above all very soft waters, concerning maintenance in an aquarium - even much more for reproduction! - a water with a pH below 6.0 and an as far as possible low total hardness (less than 80 µS) is indispensable. For reproduction of these presently still very rare specimens as well as for their common maintenance, breeders recommend a spacious tank with a length of about 100 cm. The bottom should not consist of the commonly used aquarium gravel (this one can hardly be found in the Amazon area!), but of

S03487-4 *Apistogramma elizabethae*, female.

fine white sand (no coral sand containing lime!). Tree roots arranges the aquarium in a similar way as the fish are used to in their natural habitat.Concerning the nutrition of the specimens it is conspicuous that they prefer *Crustaceae*, among these also *Artemia* nauplii as well as bigger specimens (then mostly deep frozen) of this shrimp. Basically, reproduction does not happen in a different way to other *Apistogramma* relatives.

Apistogramma eunotus Kullander, 1981

Meaning of the specific name (etymology): A composed substantival form of the Greek *eu* = strong/good and *notus* = Back; here referring to the relatively deep body shape of the specimens: with a high back.

First description: Bonner zool. Beitr. 32 (1-2), pages 183-194.

Synonyms: Before their scientific description, specimens of this species were called by different names: *Mesops taeniatus* Cope, 1870 (»near Pebas, Ecuador«); *Apistogramma amoenus* Regan, 1913 (»River Ucayali, Peru«); *Apistogramma amoenus* Kullander, 1979 (compared to *A. nijsseni*, Rio Ucayali) and others.

Distribution: From the Rio Ucayali (Pucallpa [road to Lima] downstream [road Jenaro Herrera-Colonia Angamos] as well as in the drainages of the Rio Pacaya, Rio Tigre, Rio Itaya to the surroundings of the city of Iquitos (Rio Nanay drainage near Mishana) into the Amazon, the system of the Rio Ampiyacu (Secarita del Tuyé), the lower Rio Yavarí (San Sebastian, Caño do Comprido) up to Colombia (Leticia, directly near Tsaliki's Station) and to Brazil (Tabatinga) and even in the drainage of the Rio Japurá (Paraná do Amanã, Lago Amanã) (cp. map page 43).

Partly occurs sympatrically with *A. cacatuoides*.

Habitat: It can be quite different in water quality (cp. *A. cacatuoides*), as the water coming from the Andes in the area of the central Rio Ucayali shows data that is a little bit higher than that of the Amazon lowland. Therefore, the specimens also occur in various water types (white, black and clearwater) as well as in zones with mixed water. There they also penetrate into zones with stronger current, where normally Cichlids are not found. In most habitats, at least the bottom near the banks is covered with foliage, branches and twigs, a layer that offers sufficient shelter.

S03505-3 *Apistogramma eunotus*, wild male.

S03505-3 *Apistogramma eunotus*, wild female with young.

SD and colouration: The basic colouration of male specimens is between a bright sand colour and a beige grey. Every scale is rounded dark, by which a network-like pattern covers the body. Depending on the mood, a dark horizontal band can be shown which ends in an angular caudal blotch behind a narrow bright vertical band at the end of the caudal peduncle. The beginning of the pectoral fin can be detected by an orange colour. Above the cheeks, a pattern of sky blue stripes, spots and lines can be seen. In the males, the soft ends of the dorsal and anal fin are extended and end with tips. Dorsal, caudal, anal and ventral fins shimmer bluish, at their ends brownish to weak rusty red tones can be seen which spread even more in the rounded caudal fin.

Total length: Male 8.5 cm, female 5.5 cm.

Maintenance in the aquarium and reproduction: Compared to the habitat described here, the maintenance of the deep bodied adaptive Dwarf Cichlid is normally problem-free. Reproduction follows the preceding procedure described for *A. cacatuoides*.

Apistogramma geisleri MEINKEN, 1971

Meaning of the specific name (etymology): Dedication name in honour of the well known German biologist, limnologist and fish expert Prof Dr Rolf Geisler who collected the first specimens (1967) and made three of these specimens available to the author of the first description (MEINKEN).

First description: Senckenbergiana biol. 52 (1/2), pages 35-40.

Synonyms: None.

Biotope: Rio Nanai,
near Santa Clara, near to the city of Iquitos/Perú
Date: 11.3.1987
Time: 14.15 o'clock
Water movement: Slowly flowing
Water depth: Not established
Water colour: Light tea colour

Laboratory results:

pH-value	5.26
Electrical conductivity (25°C), µS/cm	6
Acid capacity up to pH 8.2 (p-value), mmol/l	n.p.
Acid capacity up to pH 4.3 (m-value), mmol/l	0.05
As carbonate hardness, °d	0.14
Total Alkalis (Ca^{2+}/Mg^{2+}), mmol/l	0.01
As total hardness, °d	0.06
Sodium (Na^+), mg/l	2.5
Potassium (K^+), mg/l	0.55
Calcium (Ca^{2+}), mg/l	0.27
Magnesium (Mg^{2+}), mg/l	0.083
Total Iron ($Fe^{2+/3+}$), mg/l	0.36
Manganese (Mn^{2+}), mg/l	0.013
Ammonium (NH_4^+), mg/l	0.02
Nitrite (NO_2^-), mg/l	n.p.
Nitrate (NO_3^-), mg/l	n.p.
Chloride (Cl^-), mg/l	2
Sulphate (SO_4^{2-}), mg/l	3
Sodium (Na^+), mg/l	2.5
Potassium (K^+), mg/l	0.55
Total Phosphate (PO_4^{3-}), mg/l	0.04
Hydrogen Carbonate (HCO3-), mg/l	2
Oxidation	
a) As Potassium Permanganate $KMnO_4$ consume, mg/l	11
b) As Oxygen (O_2), mg/l	2.8

n.p. = not provable

S03520-3 *Apistogramma geisleri*, wild male.

Distribution: Only known from the drainage of the Rio Curuçambá near the town of Óbidos at the left (northern) bank of the Amazon northwest (opposite) of the city of Santarém in the Brazilian federal state Pará (cp. map page 59).

Habitat: In this area, the Amazon water, which is rich in sediments and troubled by muddy particles, is influenced by the entering water of the big dark Rio Trombetas. The mentioned Rio Curuçambá enters a few kilometers downstream Óbidos - coming from northwest. It comes from an area which is also crossed by the Rio Trombetas; its water is soft and very acidic. Also in this river, foliage deposits, twigs and stronger branches cover the territories of the Dwarf Cichlids near the banks. At other places, vegetation hanging over from the banks can offer shelter to the fish.

SD and colouration: The specimens show no pronounced sexual dimorphism, which means that male and female specimens are quite similar in their general appearance. Only the difference in size and the then relatively moderate colouration of the females allows us to make clear distinctions. Depending on the mood, the band on the cheeks and the snout as well as the horizontal

Biotope: Rio Trombetas,
above Oriximiná
Date: 25.10.1985
Time: 14.50 o'clock
Water movement: Sluggish near the banks, stronger in central parts of the river
Water depth: Not established
Water colour: Tea coloured

Field measurements:

pH-value (25.10., 14.50 o'clock)	5.0
pH-value (26.10., 5.20 o'clock)	5.6
Electrical conductivity (31°C), µS/cm	18
Water temp. (25.10., 14.50 o'clock)	31°C
Water temp. (26.10., 5.20 o'clock)	30°C
Air temperature	32°C

Laboratory results:

Electrical conductivity (25°C), µS/cm	20
Acid capacity up to pH 8.2 (p-value), mmol/l	n.p.[1]
Acid capacity up to pH 4.3 (m-value), mmol/l	0.40
As carbonate hardness, °d	1.1
Total Alkalis (Ca^{2+}/Mg^{2+}), mmol/l	0.02
As total hardness, °d	0.1
Calcium (Ca^{2+}), mg/l	0.29
Magnesium (Mg^{2+}), mg/l	0.21
Total Iron ($Fe^{2+/3+}$), mg/l	0.33
Manganese (Mn^{2+}), mg/l	n.p.[2]
Ammonium (NH_4^+), mg/l	0.05
Nitrite (NO_2^-), mg/l	0.06
Nitrate (NO_3^-), mg/l	n.p.[3]
Chloride (Cl^-), mg/l	2
Sulphate (SO_4^{2-}), mg/l	3
Total Phosphate (PO_4^{3-}), mg/l	0.05
Hydrogen Carbonate (HCO_3^-), mg/l	24
Carbonate (CO_3^{2-}), mg/l	n.p.[4]
Oxidation Potassium Permanganate $KMnO_4$ consume, mg/l	26

[1] not provable, determination limit 0.05 mg/l
[2] not provable, determination limit 0.02 mg/l
[3] not provable, determination limit 1.00 mg/l
[4] not provable, determination limit 2.00 mg/l

S03525-3 *Apistogramma gephyra*, wild male, blue variant.

band are shown frequently. The latter is partly visible only discontinuously and ends before a vertical caudal blotch which is situated behind a bright zone at the end of the caudal peduncle. A mostly bright blue pattern of bands, stripes and spots is conspicuous which on a blood red base extends all over the males' cheeks. In females, it is pronounced only very weakly.

Total length: Male 7 to 8 cm, female 5 to 6 cm.

Maintenance in the aquarium and reproduction: Although the habitats are far away from each other, nevertheless for this species a similar recommendation is valid as for *A. elizabethae*. For maintenance and breeding, an aquarium with a length of 60 to 80 cm is enough.

Apistogramma gephyra KULLANDER, 1980

Meaning of the specific name (etymology): From the Greek: ›*gephyra*‹ = bridge. Therewith, it shall be referred to the »superficially morphological position« of the species between *A. agassizii* and *A. pertensis* which both belong to different groups.

First description: Bonner zool. Monograph. (14), pages 131-134.

Synonyms: None.

Distribution: Hitherto known from a distribution range between the Anavilhanas Island Group in the Rio Negro, about 80 to 120 km upstream the city of Manáus in the Brazilian federal state of Amazonas to the surroundings of the city of Santarém in the federal state of Pará (Lago Jurucuí) (cp. map page 41).

Habitat: Most of the specimens which are imported, accidentally come from areas on the left side of the Amazon. The waters known here - as far as they have no direct and passing through

S03525-3 *Apistogramma* cf. *gephyra*, ›Santarém‹, white-blue variant.

S03530-4 *Apistogramma gephyra*, wild male, yellow variant.

connection with the white water river troubled by sediments - have black or at least mixed water. As shown by the water sample included here (cp. also water sample ›Anavilhanas‹ in *A. paucisquamis*), a total hardness can hardly be measured (conductivity only µS!) and the pH value of 6.4 respectively 5.82 (Anavilhanas) is in the acidic to very acidic range. Findings in the clear Lago Jurucuí, not far from the city of Santarém, prove that the specimens can also live in unmixed clear water.

SD and colouration: The species is often erroneously seen as a variety of *A. agassizii* by local fishermen and it has to be admitted that the natives are not that wrong because the lancet like caudal fin in itself can already cause this confusion. The choice of the specific name shows that the describer (KULLANDER) was thinking about the relationship. It also refers to the family closeness to the species *A. pertensis*, with whose representatives the specimens are found together in the Anavilhanas area in the Rio Negro.

On a green beige bottom - particularly in the anterior body and fin part - there is an intense blue shimmer. A cheek and snout band is often clearly visible. The lateral or horizontal band can be of different widths, depending on the origin, and is then shown intensely depending on the mood. In the posterior part part of the anterior half (in specimens of both sexes) there are two black blotches on each side of the body. In well kept females, the blotches occur particularly intensely during the breeding season.

In spring 1996, imports of a mixture from *A. pertensis* and *A. gephyra* came to Germany. The imported specimens of *A. gephyra* showed a very different appearance, sometimes extremely

Biotope: Water before Silves island, in the delta area of the Rio Urubú, peripheral blackwater lagoon not far from Silves island.
Date: 12.10.1986
Time: 14.00 o'clock
Water movement: moderate
Waterdepth: Very different, but not too deep
Water colour: Dark, sample was tea coloured

Field measurements:	
pH-value	6.420
Oxygen (O_2)	(elect.) 6.2
Water temperature	31.6°C
Air temperature	32.5°C
Laboratory results:	
Electrical conductivity (25°C), µS/cm	16
Acid capacity up to pH 8.2 (p-value), mmol/l	<0.05
Acid capacity up to pH 4.3 (m-value), mmol/l	0.16
As carbonate hardness, °d	0.5
Total Alkalis (Ca^{2+}/Mg^{2+}), mmol/l	0.05
As total hardness, °d	0.3
Calcium (Ca^{2+}), mg/l	0.67
Magnesium (Mg^{2+}), mg/l	0.28
Total Iron ($Fe^{2+/3+}$), mg/l	0.03
Manganese (Mn^{2+}), mg/l	<0.02
Ammonium (NH_4^+), mg/l	0.07
Nitrite (NO_2^-), mg/l	<0.02
Nitrate (NO_3^-), mg/l	<1
Chloride (Cl^-), mg/l	3
Sulphate (SO_4^{2-}), mg/l	8
Total Phosphate (PO_4^{3-}), mg/l	0.05
Hydrogen Carbonate (HCO_3^-), mg/l	10
Oxidation Potassium Permanganate $KMnO_4$, mg/l	324

different from each other. Some particularly well developed males in fact, showed themselves very simular to *A. pertensis* in body and fin shape, but the colouration reminded one of *A. agassizii*. While looking at such specimens, one is automatically reminded of the meaning of the specific name (= bridge, see above).

Total length: Male 6 cm, female 4.5 to 5 cm.

Maintenance in the aquarium and reproduction: Similar to that given for *A. agassizii*.

Apistogramma gibbiceps MEINKEN, 1969

Meaning of the specific name (etymology): From the Latin *gibbus* = hump and *ceps*, *cephalus*, *caput* = head: ›with a hump head...‹ In one of his works according to this, MEINKEN proposed to give specimens of this species the German name ›Ramskopfcichlide‹. But this name was rejected by SCHMETTKAMP (1982) ›because of misguidance‹ (»the forehead hump is not present in living specimens and can be attributed to an insufficient preservation«). Instead, ›Schwarzbinden-Apistogramma‹ was proposed.

First description: Senckenbergiana biol. 50 (1/2), pages 91-96.

Synonyms: *A. roraimae*.

Distribution: Brazil; drainage of the Rio Negro, probable in the central part, near the inflow of the Rio Uaupés. Meanwhile the specimens described as *A. roraimae* are considered belonging to the species *A. gibbiceps*, its distribution range (Rio Branco) has to be added to the one mentioned above (cp. map page 68).

Habitat: Small waters, respectively small affluents of the Rio Negro. The specimens of this species were proved to be sympatric with *A. pertensis*, *A. uaupesi*, *A. gephyra* and *A. paucisquamis* (RÖMER, 1990).

SD and colouration: The sexes can only be distinguished with the beginning of sexual maturity. The very intense cheek stripe has the same dark colour as the horizontal band. The strong padded lips and the throat are red orange. The lower part of the body shows a beige tone and has three clear diagonally running stripes. The skins of the first three dorsal fin rays are black, on the other side those of the fourth and fifth rays are red. The entire dorsal is bluish. The skins of the ventral fins as well as the anal fin show a grey blue to orange tone. The bilobated caudal fin shows a red brown cube pattern. The dorsal and anal fin, but above all the caudal fin of the

S03535-3 *Apistogramma gibbiceps*, rivaling wild males.

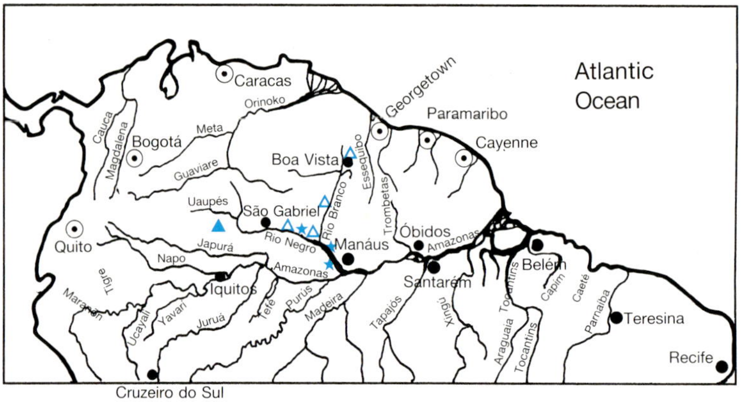

★ *A. paucisquamis*
△ *A. gibbiceps*
▲ *A. personata*

males, produce attractive filaments. In well developed males, the membranes of the third to fifth dorsal fin ray are extended a bit more.

The fins of the females are clear and do not show filaments. The caudal of young specimens is slightly truncated, in adults it is clearly truncated.

In neutral mood, in both sexes, *A. gibbiceps* shows an intense black horizontal band reaching into the caudal fin. In affection, the stripes at the lower belly are particularly dark. They are pronounced in dark colour especially under affection and therefore they are very concise. The males' cheeks can be more or less golden yellow and covered by a blue pattern.

A. gibbiceps occurs in various colour forms. In the dorsal, anal and caudal fin, the specimens which in recent years were mainly introduced occasionally with other Dwarf Cichlids, show a tender lilac red with sky blue ray like design. But obviously, there are also geographic varieties in which the lilac red is replaced by a tender yellow, more rarely by a tender blue. These pastel tones in the fin membranes only show up particularly conspicuous in state of affection. Then the specimens show themselves as tender filigree beauties.

Total length: Male 7 to 8 cm, female 5 to 6 cm.

Maintenance in the aquarium and reproduction: Until today, not much is known about reproduction. Anyway, the specimens belonging to the species are not easy to breed. As the specimens come from areas with extremely soft water, this fact should already be taken into consideration for the maintenance.

S03535-3 *Apistogramma gibbiceps*, wild male.

S03545-3 *Apistogramma gossei*, wild male, Brazilian form.

Apistogramma gossei KULLANDER, 1982

Meaning of the specific name (etymology): In honour of the Belgian ichthyologist Dr Jean-Pierre GOSSE who collected and preserved the specimens already in 1962. Another collector, the Frenchman F. D'AUBENTON, found more material in 1976.

First description: Cybium 6 (4), pages 64-72.

Synonyms: None.

Distribution: As far as known until today, the distribution ranges are exclusively in French Guyana, as well as in a narrow area of the southern neighbouring Brazilian federal state Amapá (Rivières Appouague, Oiapoque/Oyapock, Rio Araguarí). The subsequently mentioned locality ›Crique Macouria‹ is close to the coast north of the capital Cayenne (cp. map page 70).

Habitat: As one of the first aquarists, BORK examined the habitats of this Dwarf Cichlid in French Guyana (cp. BÖHME, FEY, BORK & BERGK, 1990) and brought the specimens alive back to Germany. Due to his data, there they live in small rainforest waters (streams, ditches etc.) with low current. The water is mostly clear or slightly brown, and acidic (cp. measurements). The bottom of the habitat can be covered with gravel and big stones between which mostly additional foliage and dead wood are accumulated.

SD and colouration: Cheek, snout and forehead stripes are mostly clear to see. This is also true for the lateral band which ends in a bright stripe on the caudal peduncle, where - in front of the base of the caudal fin - there is an oval vertical caudal blotch. In the lower part of the body, there are two to three stripes. The cheeks and the throat are covered by a pattern of red and blue lines and spots. The golden blotches shown on this photo cannot be seen in this form in all males. To the viewer, they have the effect of the egg spots of African Cichlids, but at this part,

Biotope: French Guayana,
spring stream of Crique Macouria, near the town with the same name
Date: 12.8.1989
Time: 11.00 o'clock
Water movement: Slow
Water depth: 0.2 to 1 meter
Water colour: At 26°C amber coloured

Laboratory results:

pH-value	5.60
Electrical conductivity (25°C), µS/cm	20
Acid capacity up to pH 8.2 (p-value), mmol/l	n.p.
Acid capacity up to pH 4.3 (m-value), mmol/l	0.05
As carbonate hardness, °d	0.14
Total Alkalis (Ca^{2+}/Mg^{2+}), mmol/l	0.03
As total hardness, °d	0.14
Sodium (Na^+), mg/l	2.6
Potassium (K^+), mg/l	0.31
Calcium (Ca^{2+}), mg/l	0.26
Magnesium (Mg^{2+}), mg/l	0.44
Total Iron ($Fe^{2+/3+}$), mg/l	1.5
Manganese (Mn^{2+}), mg/l	0.005
Ammonium (NH_4^+), mg/l	0.02
Nitrite (NO_2^-), mg/l	<0.02
Nitrate (NO_3^-), mg/l	<1
Chloride (CL^-), mg/l	4
Sulphate (SO_4^{2-}), mg/l	<1
Total Phosphate (PO_4^{3-}), mg/l	0.11
Hydrogen Carbonate (HCO_3^-), mg/l	3
Oxidation a) As Potassium Permanganate $KMnO_4$ consume, mg/l	14
b) As Oxygen (O_2), mg/l	3.5

n.p. = not provable

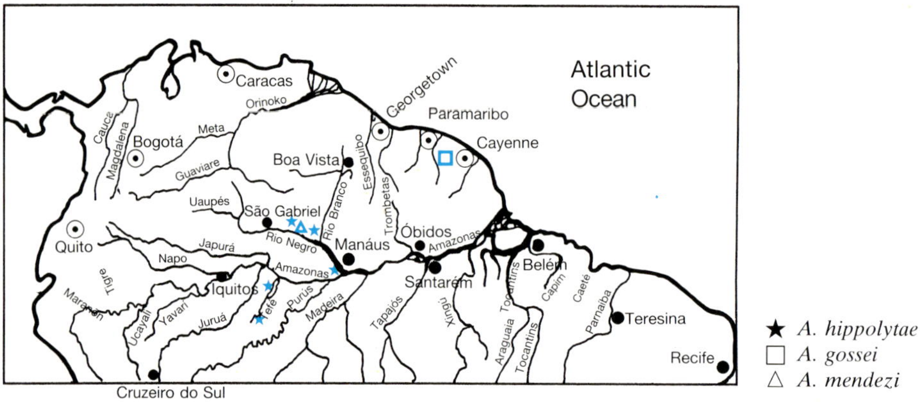

most of the specimens have a pattern which consists of red and blue tones, just like the rest of the anal and dorsal fin. The truncated caudal fin of the males is clear to milky white. During the breeding season, female specimens become yolk yellow and develop deep black markings (cheek and horizontal bands, caudal blotch). The anterior rays of the dorsal and ventral fins are also coloured deep black.

Total length: Male 6 cm, females 4 cm.

Maintenance in the aquarium and reproduction: Although in their natural habitat the specimens live in very soft and slightly acidic water, their maintenance in the aquarium at a water hardness of 12°d and a pH value of 7.0 did not cause any problems. For the first time, Böhme (1989) and Machnik (1993) succeeded in reproduction of wild fish each time. Moreover, *A. gossei* also reproduced further at a pH value of 7.0 and a total hardness of 6°d without any problems.

Apistogramma guttata Antonio, Kullander & Lasso, 1989

Meaning of the specific name (etymology): Adjectival form adapted to the female generic gender of the Latin wod *guttatus* = flecked, here referring to the pattern, particularly in males.

First description: Acta Biol. Venez. 12 (3-4), pages 131-139.

Synonyms: None.

Distribution: As type locality, the system of the almost 200 km long Rio Moichal Largo in the north eastern Venezuela is given, which can be comfortably reached from the tourist centres around the coastal cities of Barcelona/Puerto la Cruz on the way to the Orinoco delta between the towns of Maturím and Tucupitá. The river flows via the Caño Mánamo into the southwestern Golfo de Paría between the mainland and the Island of Trinidad (cp. map page 73).

Habitat: The bottom near the banks of the river consist of fine sand, but it is partly covered with a thick layer of mud. During daytime, the slightly tea coloured water shows mostly temperatures from 27 to 28°C and can at noon heat up to more than 30°C. Additionally, the water is very soft (20 to 50 µS) and its pH value of 5.3 to 5.5 is very acidic.

Various aquatic and swimming plants border the areas with shallow water, above all, *Cabomba aquatica* and various *Echinodorus* species. At the surface, there is mainly *Eichhornia azurea* and *E. crassipes*. The authors (Antonio et al.) report about catches of *Apistogramma* in ›swimming meadows‹ formed by *Paspalum repens*. On the base of our own collections, this species could not be proved in these vegetation carpets, but it was present in the shallow water near the vegetation present there. In the pasture ponds along the previously mentioned road which are mainly bordered by palm trees, we also found *Microgeophagus ramirezi*.

S03550-4 *Apistogramma guttata*, male.

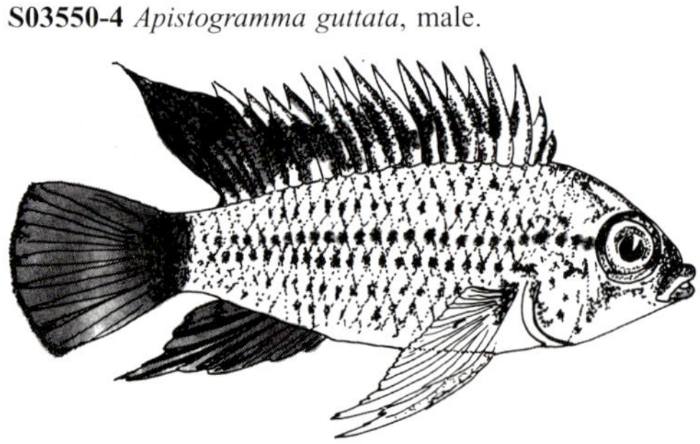

SD and colouration: Adapted to the sandy bottom of their natural habitat, the male specimens of *A. guttata* have a sand beige base colouration. Females are coloured similarly, but they stay brighter and change their body colouration only during the breeding period, into a brass like yellow above which certain markings (cheek and horizontal band, caudal blotch) are conspicuous. The two anterior shorter dorsal fin rays are also coloured black.

In the males the common black markings, but particularly the horizontal band, seem mostly to be dissolved in spots, just like the flanks of the fishes are covered with five rows of black spots. The fins show no conspicuous colouration and are almost colourless. In adult specimens, only at the upper and lower edge of the truncated caudal fin can tender red tones be seen, which by the way are also known from specimens of *A. hoignei* which were collected not far away on the border of the Orinoco delta.

Remarks: In some shipments from 1996 which were declared as ›*A. guttata*‹ from Venezuela, with a 100% certainty there were young specimens of *A.* spec. ›Breitband‹ which could not yet bet differentiated to their sex. Indeed, these could not come from the drainage of the Rio Morechal north of the Orinoco delta, but from an area around Puerto Ayacucho at the central Orinoco.

Total length: Male 6 cm, female 3.5 cm.

Maintenance in the aquarium and reproduction: The fish turned out to be relatively adaptive. They can be kept at the common water temperature of 27°C in soft water with a pH value near neutral (7.0) without problems, and besides the usually given deep frozen and live food, they occasionally accept even flake food.

But if one wants to make the fish spawn, the water has to be adapted to the conditions in the natural habitat: It should be very soft and very acidic. Moreover, the breeding aquarium which is not too small (length more than 60cm) should have a dense aquatic vegetation. Reproduction itself follows the principles of the male-mother-family, whereby the female alone takes care of the fry, while the partner protects the territory. The eggs deposited in a cave (about 100) develop in about 4 days to the hatching of the larvae and after five more days the young fish swim free. They immediately consume *Artemia* nauplii.

Apistogramma hippolytae Kullander, 1982

Meaning of the specific name (etymology): A wordplay taken from Greek mythology: To the honour of Hippolyte, the ›Queen of the Amazones‹. During the rubber boom, the centrally positioned city of Manáus, situated in the Brazilian federal state Amazonia, from where many Dwarf Cichlids come to us, was seen as the ›Queen of Amazonia‹.

First description: DCG-Informationen 13 (10), pages 181-193.

Synonyms: None.

Distribution: While Kullander could ›only‹ name the system of the Rio Manacapurú and areas in the system of the Rio Negro (›central lake on the Ilha de Buiu-açu, near the Rio Urubaxi‹ and ›central lake of an island downstream the Rio Daraá‹), Koslowski (1994 a) also reports about the existence of specimens of this species in the Rio Tefé, whereby the knowledge about the distribution of this Dwarf Cichlid was considerably widened. All mentioned habitats are black water territories (cp. map page 70).

Biotope: Lago Manacapurú,
the lowest in this chain of lakes
Date: 3.10.1986
Time: 7.00 o'clock
Water movement: sluggishly flowing
Water depth: Not established
Water colour: Light tea brown

Field measurements:	
pH-value	6.76_{20}
Water temperature	29.0°C
Air temperature	28.0°C

Laboratory results:	
Electrical conductivity (25°C). µS/cm	40
Acid capacity up to pH 8.2 (p-value), mmol/l	<0.05
Acid capacity up to pH 4.3 (m-value), mmol/l	0.26
As carbonate hardness, °d	0.7
Total Alkalis (Ca^{2+} + Mg^{2+}), mmol/l	0.09
As total hardness, °d	0.5
Calcium (Ca^{2+}), mg/l	2.9
Magnesium (Mg^{2+}), mg/l	0.53
Total Iron ($Fe^{2+/3+}$), mg/l	0.15
Manganese (Mn^{2+}), mg/l	<0.02
Ammonium (NH_4^+), mg/l	0.07
Nitrite (NO_2^-), mg/l	<0.02
Nitrate (NO_3^-), mg/l	<1
Chloride (Cl^-), mg/l	3
Sulphate (SO_4^{2-}), mg/l	9
Total Phosphate (PO_4^{3-}), mg/l	0.05
Hydrogen Carbonate (HCO_3^-), mg/l	16
Oxidation $KMnO_4$ consume, mg/l	325

S03560-3 *Apistogramma hippolytae*, wild male

Habitat: The waters mentioned show a total hardness of less than 1°d, a differently acidic pH value as well as an extremely low nitrate value (<1). In the more shallow parts close to the banks, the sandy bottom above which the tea coloured water flows is covered with organic waste of the surrounding forest, which means foliage, twigs and coarse branches. Aquatic plants can not grow in the water: It is too dark to allow the plants a regular photosynthesis.

SD and colouration: With such a widely scattered dissemination as the above mentioned it was inevitable, that in colouration and pattern of the caudal fin there are quite some (more or less fine) differences. So the colouration of the males can differ in some details (for example in the colouration of the narrow submarginal dorsal band or in the size and form of the caudal blotch). Compared to *A. rupununi* which is similar in the lateral blotch and before its revalidization by KULLANDER (1989) was considered ›Zweifleck-Apistogramma‹ and possible synonym of *A. steindachneri*, it has to be stated that contrary to those of *A. rupununi* the males of *A. hippolytae* show a rounded caudal fin (in *A. r.* it is truncated). Additionally, in *A. rupununi* the lateral blotch is narrower and reaches only up to the upper lateral line instead of (like in *A. hippolytae*) to the base of the dorsal fin (cp. photos!). In their breeding colouration, females show yolk yellow tones which extend into the fins. In this stadium, the cheek stripe as well as the lateral and caudal blotch are marked just as deep black as the anterior dorsal fin ray.

Total length: Male 7cm, female 5cm.

Maintenance in the aquarium and reproduction: For maintenance and breeding, these Dwarf Cichlids from central Amazonia need very soft and very acidic water which is absolutely free from nitrate (cp. water tests). A tank should be at least 60 cm long and a partial water change once a week should be done. The socialization with small and relatively peaceful fish can be as helpful as a well growing aquatic vegetation. These fish can raise 160 to 200 offspring. The breeding colouration mentioned before can be seen mostly after the reddish eggs deposited with one pole have developed far enough so that the hatching of the larvae (after about four days) is near. Now the mother shows her most beautiful dress and after five more days she leads the fry around in the tank for their first ingestion.

S03569-4 *Apistogramma hoignei*, wild male, Orinoco delta.

Apistogramma hoignei MEINKEN, 1965

Meaning of the specific name (etymology): Dedication name in honour of the collector Leo HOIGNE who in 1964, at first gave the two type specimens (not a pair, but two females as KULLANDER stated in a later examination) to the then very active American editor Albert KLEE (›The Aquarium‹), who later however forwarded them to the author (MEINKEN) as preserved material.

First description: Senckenbergiana biol. 46 (4), pages 257-263.

Synonyms: None.

Distribution: The preserved material which was to MEINKEN's disposition for the first description comes from affluents of the Camaguan Swamps at the lower Rio Portuguesa (Venezuela). It flows into the Rio Apure. The area is northwest of the capital of the state of Guáricoin, San Fernando de Apure. Upstream of this city, the two rivers unite and the Rio Apure finally leads its water via the Rio Orinoco into the Atlantic Ocean.

During our own collecting trips to the Orinoco delta, in bigger ponds often surrounded by palm trees (among those also watering places), specimens could be caught which are morphologically very similar to the species, but they partly showed more intense colours. Remarks by KULLANDER (pers. comm.) say that these specimens are members of the species *A. hoignei* (cp. map below).

Habitat: In these related habitats (cp. water measurements page 34) one knows the Cichlids from shallow parts where they live with relatively few demands. If in border areas of slowly flowing waters or in bigger ponds: The number one priority of the specimens is their safety.

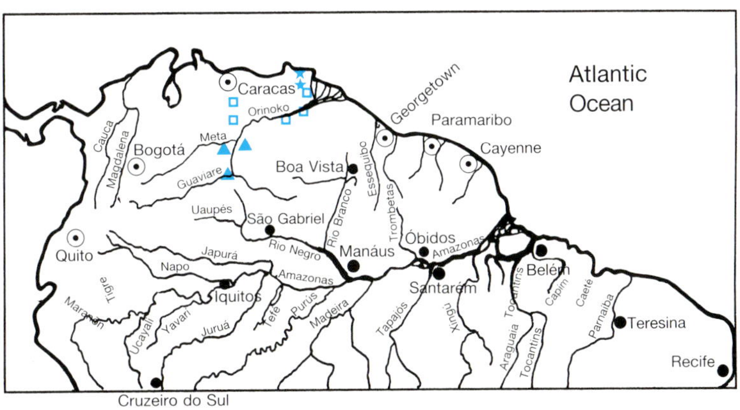

★ *A. guttata*
□ *A. hoignei* (+ *spec. „Caura“*)
▲ *A. hongsloi*

Hiding places are found under twigs, in coarser branches and foliage, as well as under the overhanging vegetation of the banks. As in many areas cattle is raised, the natural water conditions can be negatively influenced by agricultural factors.

Remarks: Closely related specimens (or even of the same species?) were found in the Rio Caura which flows from the south into the Orinoco, as well as in the western drainage of the Orinoco delta (cp. *A.* spec. ›Rio Caura‹).

SD and colouration: The opulent fins of the hefty males are conspicuous. Compared to specimens from the Portuguesa-Apure area, certain males (not all because one finds purely grey specimens as well as intensely coloured ones in the same habitat) from the Orinoco delta show an intense blue proportion all over the posterior part of the body, as well as on the dorsal and anal fin. In the more colourful males, above all the more intense golden yellow tone on the forehead and the neck, as well as on the posterior parts of the unpaired fins, is conspicuous. Particularly in specimens from the Rio Portuguesa drainage, the upper and lower rays of the caudal fin show red tones. Its central part is transparent, its posterior edge is bilobated when the specimens are adult.

In their wedding dress and while caring for the fry, the females of this species also show a yellow colouration. Then the two to three anterior dorsal fin rays and the anterior ventral fin rays are deep black. Additionally, stripes on the cheeks and the neck as well as the lateral band with the caudal blotch are shown in intense black colouration.

Total length: Male 7 cm, female 5 cm.

Maintenance in the aquarium and reproduction: Maintenance and breeding in the aquarium do not cause great problems as long as the water is as soft as possible and above all nitrate free. The pH value should be in the acidic range. With a normal temperature of 27°C the larvae have an incubation time of about 4 days up to hatching, and the young fish swim free after five more days so that they can independently take food.

Apistogramma hongsloi KULLANDER, 1979

Meaning of the specific name (etymology): Dedication name in honour of the Swedish collector Thörborn HONGSLO who besides this one, discovered further species (*A. iniridae*, *A. macmasteri* and *A. viejita*).

First description: Zool. Scripta 8, pages 71, 74-75.

Synonyms: None.

Distribution: The type material collected by HONGSLO comes from the estuary areas of the Rio Vichada and Rio Guarrojo as well as from the Caño Perro which is 155 km northnortheast and pours its water into the Rio Meta. The com-

S03589-4 *Apistogramma hongsloi*, impressing male, blue.

S03592-5 *Apistogramma hongsloi*, pair.

plete distribution range can be seen in the Llanos of the Colombian Oriente, the tropical lowland in the eastern part of the country. The former steppe landscape is nowadays more and more used for agriculture, mainly for cattle raising. In a similar way, the neighbours in Venezuela utilize the northerly attached area in the Rio Meta and Rio Apure drainages where the distribution range of *A. hoignei* follows (cp. map page 73).

Habitat: The water in the drainage of the central Rio Orinoco is very soft (total hardness <1°d) and additionally very acidic (pH value 6.0 to 5.4). Here the Dwarf Cichlids live in the shallow zones near the banks of the rivers and in the lake-like bulges. Furthermore they can be found in ditches and gullies. The water temperature depends on the season of the year, the same is true for the water level. Where there are no trees along the banks, the overhanging vegetation is used for shelter.

SD and colouration: Depending on the origin in the extended distribution range, the colouration of the males can vary considerably and is then qualified by aquarists as ›very beautiful‹, ›beautiful‹ or ›colourless‹. The Colombian specimens already introduced in the seventies and called ›Rotstrich-Apistogramma‹ belong to the category ›very beautiful‹ and still today, offspring specimens are purchased at a high pricc. Thcy have a golden yellow head, breast and throat. The posterior part of the body as well as the dorsal, anal and pectoral fin are silvery grey and shimmer slightly bluish. Around the posterior caudal peduncle there is a narrow bloodred band whose extensions even reach into the base of the caudal fin. A red line runs from the lower part of the band along the lower part of the body into the anal area.

The beautiful colouration of these specimens did not leave a number of breeders in the eastern part of Germany calm, in recent years their import rate was strictly limited, and they tried to strengthen the specimens' colour, whereby really splendid males occured. In the so-called Farbform I, an early phase of development can be seen, Farbform II is the actual highest phase with a high red percentage, particularly in the anal area.

Females have a yellowish base tone which is intensified during breeding.

Total length: Male 8 cm, female 5 cm.

S03592-5 *Apistogramma hongsloi*, male in neutral mood, red.

Maintenance in the aquarium and reproduction: Unfortunately, there are but a few imported specimens. Most of the specimens offered by commercial trade come from broods, but therefore are not less beautiful! The quite big and deep bodied specimens are kept in aquaria with a length of more than 80 cm. For the maintenance of adapted specimens water with 6 to 10°dGH and a slightly acidic pH is sufficient. For reproduction, for one male several (two to three) fe-males are added. For every specimen, an appropriate (cave-like) hiding place has to be prepared. Now the water has to be as soft (up to 3°dGH) and as poor in nitrate as possible. The pH value has to be acidic (below 6.0). The water temperature is adjusted to the common 27°C. A rich but above all, very variable diet goes without saying. Whether ›Rotstrich-Apistogramma‹ or pedigree, the male will soon shine in the most beautiful colours and court the present partners. The breeding dress of the females shows less golden yellow, rather sand or lemon yellow tones which seem to be overlapped by a green tone. Their care instinct is very much pronounced which can particularly be seen when some other fish are socialized in the aquarium and disturb the future mother. She occasionally tries to earn respect by starting short attacks.

The development of the eggs depends on the temperature and - as usual - at 27°C the larvae take about four days up to hatching and until they swim free, five more days pass. As the development from the egg to the young fish happens in the mother's cave, some aquarists may be surprised when one day the female proudly shows off her offspring. The young immediately consume *Artemia* nauplii. They develop quickly, but one can soon see that the development does not advance identically in all specimens. There are often certain males that grow faster and then soon they show a certain dominant behaviour pattern. At first they can be detected by a red blotch on the caudal peduncle. With the help of showing traits of the female colour pattern, weaker males try to avoid the pressure of the dominant males.

Apistogramma inconspicua KULLANDER, 1982

Meaning of the specific name (etymology): Adjective form of the Latin ›inconspicuus‹ = ›not very conspicuous‹, adapted to the female generic gender. Here the name refers to the ventral stripes which for example in the closely related species *A. commbrae* are much more intensely pronounced.

First description: Zool. Scripta 11 (4), pages 307-313.

Synonyms: None.

Remarks: Specimens of this species that originate from the distribution range of *A. commbrae* and *A. trifasciata* were collected a long time before the scientific description (HASEMAN, 1909),

S03600-4 *Apistogramma inconspicua*, impressing male.

but the preserved material was classified to the two species mentioned. It was a later collecting trip by the Belgian ichthyologist Jean-Pierre Gosse in 1977 that brought further material and enabled a definition and a scientific description of the new species.

Distribution: The type material comes from a tributary of the Rio San Joaquín in the Guaporé drainage in Bolivia. On the one side, the distribution range of this species reaches through the watershed to the Rio Amazonas and on the other side to the Rio Paraná. The Rio Guaporé in the Mato Grosso still belongs to the Amazon drainage. Against it, the nearby Rio Paraguay (originating also north of the Pantanal but further to the east in the Brazilian federal state of Mato Grosso) leads its water to the south, via the Rio Paraná into the southern Atlantic Ocean. Therefore, *A. inconspicua* indeed occurs in the drainage of the Rio Guaporé as well as in the upper Rio Paraguay (cp. map below).

Habitat: The living conditions for the specimens can be different as rainforest areas dominate (even when ›in retreat‹), but secondly also a savannah like landscape is spreading which has been caused by burning down the jungle for the purpose of agriculture (cp. water samples page 127 and 161).

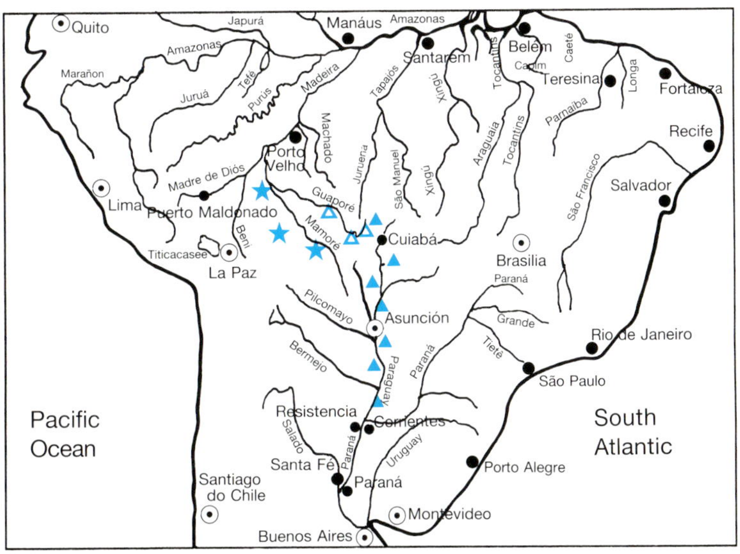

★ *A. linkei*
□ *A. inconspicua*
▲ *A. commbrae*

SD and colouration: The close relationship of the specimens to *A. commbrae* is unmistakable - as far as it concerns the optical appearance: The representatives of both species have a slightly rounded caudal fin on which a pattern of rust red spots arranged in arched lines can be seen. In its basal part, the not particularly high dorsal fin is dark, almost black. Its dark seam can only be seen during the breeding season when the males change their colouration and take a much more beautiful ›dress‹. The ends of the dorsal and anal fin have tips and the mentioned pattern of the caudal fin extends into the posterior parts of the dorsal and anal fin. On the lower part of the head and the gills there is a pattern of sky blue spots and lines (as it can also be seen in *A. commbrae*). The mostly irregularly running cheek and lateral band which in the normal dress can be seen, ends in a big wide blotch on the caudal peduncle. If while breeding the splendid colouration is shown, these bands are suppressed together with the blotch on the caudal peduncle.

Female specimens are pale yellow, this is a colour that is intensified in the breeding season. Then also the two anterior dorsal fin rays as well as the anterior ray of the ventral fins and the mentioned bands are deep black; the same is true for the blotch on the caudal peduncle.

Total length: Male 7 cm, female 4.5 cm.

Maintenance in the aquarium and reproduction: Keeping and reproducing specimens of this area turn out to be less demanding, compared to their relatives from central Amazonia. They require - even for reproduction - a water poor in nitrate, acidic in pH value but it does not have to be soft. As disclosed by GRAD in a report in DCG-Info in 1986, *A. inconspicua* repeatedly reproduced in a 450 liter aquarium (socialized with *Laetacara dorsigera* as well as with some Tetras and *Corydoras*) at a water temperature of 25°C, a total hardness of 21°d and a pH value of 8.5(!). Under the conditions named before, the development of the fry, under the temperature mentioned, three to four days to hatching of the larvae and about five more days until swimming free.

Apistogramma iniridae KULLANDER, 1979

Meaning of the specific name (etymology): To the Rio Inirida in the east of Colombia, a tributary with mixed water near the Puerto Inirida

S03605-3 *Apistogramma iniridae*, threatening male.

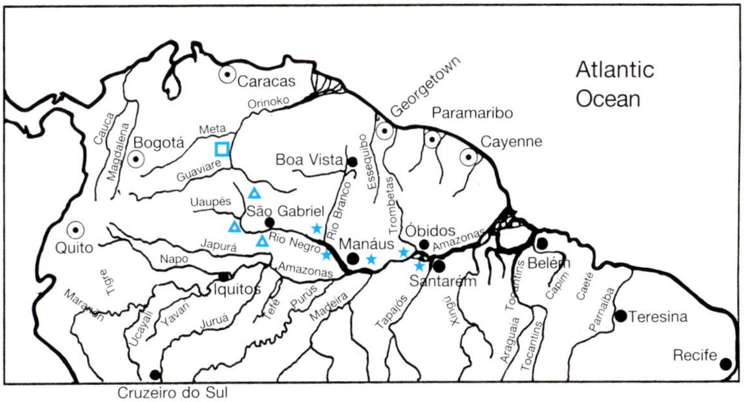

★ *A. pertensis*
□ *A. iniridae*
△ *A. meinkeni*

directly flowing into the Rio Guaviare which already a few kilometers further pours its water into the Rio Orinoco via the quite black Rio Atabapo. By the way, in the headwater area of the Rio Atabapo, there is the watershed that separates the Atabapo and the Rio Negro and lets them flow into the according directions.

First description: Zool. Scripta 8, pages 76-79.

Synonyms: None.

Distribution: As far as known today, the distribution ranges are exclusively in the drainage of the Rio Inirida in the Oriente of Colombia. The type material comes from the densely vegetated banks of the Caño Bacón, a small stream to the Rio Inirida near Pueblo Remanso where the Swedish bacteriologist Thorbjörn Hongslo found the specimens together with *Paracheirodon axelrodi* and *Dicrossus filamentosus* (cp. map above). Already in 1975, Mayland could catch some specimens in the same area but without thinking at that time about the nomenclatoric affiliation of these Dwarf Cichlids.

Habitat: In the area of the prospering border town of Puerto Inirida, where for the inhabitants the catch of eaten fish and ornamental fish has absolute priority, the water is extremely soft and the pH value is very acidic. For a long time, the original rainforest has been cut in vast areas. Nevertheless, for the fish in the habitat there are enough hiding places. The bottom near the banks is covered with fine and coarse sand, just like smaller border waters and streams like the one mentioned permit the assumption of accumulations of branches and foliage.

SD and colouration: *A. iniridae* belongs to the *pertensis* group. Male specimens have a high dorsal fin and are therefore classified into this group. Their caudal fin is oval to rounded, in completely adult specimens the ventral fins are well extended. When it comes to colouration, the specimens are not particularly attractive.

On an olive grey basic tone, the males show a cheek and snout stripe. In the centre of the flanks, the horizontal band is wider and passes through a caudal blotch that is pronounced only very weakly into the caudal fin. A particular trait is conspicuous: This is a row of arranged slanting stripes on the lower ventral part showing themselves very impressive and as black rectangles depending on the mood near the lateral band. The dorsal, caudal, anal and ventral fins show blue tones with patterns or areas of rust red colouration.

Remark: Dwarf Cichlids of this species are seldom imported from Colombia under their true name. Occasionally one finds young specimens in importations of the Neon Tetra (*Paracheirodon innesi*) from Colombia (!), but even in this case they are very rare.

Total length: Male 7 cm, female 5 cm.

Maintenance in the aquarium and reproduction: The specimens are kept in tanks of more than 60 cm length - at best in pairs. The bottom should be dark, the water as soft as possible (in the Rio Inirida total hardness could not be detected and the µS indicator of my device did not even move!) and the pH value shall be at 5.0 and thus in the very acidic range. As always, it is recommended to introduce hiding places in the form of tropical wood or coconut shells. Additionally, a rich background vegetation offers the fish a feeling of safety.

Reproducing *A. iniridae* in the aquarium is generally considered difficult. The reddish eggs which are adhesive at one pole and have a diameter of almost 1 mm are deposited by the female in the cave and count from 80 to 120. After 70 to 90 hours and with a water temperature of 28°C, the larvae hatch, whereby the female is collaborative. Then they are transferred several times. Their further development up to swimming young fish takes about some more 96 to 120 hours. It is now the female who cares for the fry alone showing its most beautiful yellow colours and it attentively leads the offspring around in the aquarium for the first ingestion. In young females spawning for the very first time, it can happen that the eggs are eaten, fall victim to other cohabitants or the water quality does not allow a natural development. In the former case, one can assume that the breeding instinct is not yet developed well enough. In the latter case, the acidity of the water does not suffice. It has to be well lowered (which means below [!] the 5.0 value). Then one has to hope for another spawning.

Apistogramma juruensis KULLANDER, 1986

Meaning of the specific name (etymology): To the locality in the Rio Juruá, a bigger very long and above all very curvy tributary of the Solimões. The headwaters of the river are partly on the other side of the Peruvian border southeast of the city of Pucallpa. But the type material comes from the Brazilian part of the river in the federal state of Acre.

First description: Cichlid fishes of the Amazon drainage of Peru. Swed. Mus. Nat. Hist., Stockholm (431 pages), pages 177-181.

Synonyms: None.

Remarks: KULLANDER notes in his description: the material on which the description is based, was firstly identified as *A. cacatuoides* (upon which the author numerates further discriminating traits). The species belongs to the *cacatuoides* group.

Distribution: Hitherto only known from the first locality, a small stream in the upper Rio Juruá (at the Igarapé Formosa) near the city of Cruzeiro do Sul. A further locality has been proven in a pond beside the tributary, the Rio Tarauacá.

Right side: Flooded banks in the drainage of the Rio Juruá. Habitat of *A. juruensis*.

S03615-4 *Apistogramma juruensis*, wild male, blue variant.

S03615-4 *Apistogramma juruensis*, wild male, colourful variant.

Due to my experiences with occasionally imported specimens, it can be considered possible that the hefty little Cichlids are also distributed in the lower part of the river (cp. map page 51). Besides the Rio Purus, the Rio Juruá is the windiest among the long tributaries of the upper Amazon or the Solimões. Near the banks, the curves or loops do not only form the common kinds of slopes, but also river loops which in the course of time transform into arms that have but little water or are even dead. Then, just like in the latter case, as ›dead lagoons‹ due to the season they have a considerably changing water level.

Habitat: The specimens of this very beautiful species from the western central Amazonia are imported from the already mentioned Rio Juruá which between the Rio Jutaí and the Rio Tefé flows into the Solimões (which is the name for the Brazilian Amazon between the border area Perú/Colombia/Brazil and the entry of the Rio Negro). Here the character of the water is typical for this area: very soft and practically without measurable nitrate concentrations, but very acidic. As already said, the habitats can be different, but in most cases the bottom is sandy or covered with sediments, near the banks it is covered with dead branches ("dead wood") and foliage. Aquatic vegetation is relatively limited and can be detected mainly near the banks.

SD and colouration: In both sexes, *A. juruensis* shows much similarity to the representatives of *A. cacatuoides*. But the hefty head of the males with the padded lips reminds one of the equally related *A. norberti*. The very short cheek stripe has the same black green colour as the horizontal band which with a clear distance ends in front of the identically coloured blotch on the base of the caudal fin. Temporarily, even if very weak, a lateral blotch can be seen. The lips and the

Biotope: Upper Rio Juruá,
upstream the city of Cruzeiro do Sul, Acre (Brazil)
Date: 16.11.1996
Time: 10.00 o'clock
Water movement: Rapidly flowing
Water colour: White water with lots of sediments
Surface temperature: 26.3°C

Laboratory results:

pH-value	6.91
Electrical conductivity (25°C), µS/cm	65
Acid capacity up to pH 8.2 (p-value), mmol/l	n.p.
Acid capacity up to pH 4.3 (m-value), mmol/l	0.64
As carbonate hardness, °d	1.8
Total Alkalis, mmol/l	0.26
As total hardness, °d	1.5
Sodium (Na^+), mg/l	2.6
Potassium (K^+), mg/l	2.5
Calcium (Ca^{2+}), mg/l	8.4
Magnesium (Mg^{2+}), mg/l	1.3
Total Iron (Fe^{2+}/Fe^{3+}), mg/l	0.95
Manganese (Mn^{2+}), mg/l	0.014
Ammonium (NH_4^+), mg/l	0.13
Nitrite (NO_2^-), mg/l	<0.02
Nitrate (NO_3^-), mg/l	1.2
Chloride (Cl^-), mg/l	0.2
Sulphate (SO_4^{2-}), mg/l	2.5
Phosphate (PO_4^{3-}), mg/l	0.23
Hydrogen Carbonate (HCO_3^-), mg/l	39
Oxidation	
a) As $KMnO_4$ consume, mg/l	30
b) As Oxygen (O_2), mg/l	7.6

n.p. = not proven

S03615-4 *Apistogramma juruensis*, old male, beige variant.

S03615-4 *Apistogramma juruensis*, wild female.

throat are slightly coloured orange red. The lower part of the body shows a yellow beige tone and additionally it has three dark stripes. The membranes of the first three dorsal fin rays are black, the fourth and the seventh are a slightly orange red colour. The entire dorsal fin is iridescent bluish. The menbranes of the ventral and anal fins show a grey blue to slightly orange tone. On the bilobated caudal fin, a red brown cubic pattern can be seen.

As far as could be proven, *A. juruensis* males of different colouration occur. Besides the specimens with the colouration described before which have the English name "orange spike", there are some of the variety "black spike" whose fourth to seventh dorsal fin rays show dark grey to black colour elements. In this concern, RÖMER & SOARES (1996) were interested in the reasons based on the population biology and they came to the conclusion that males of the type "orange spike" are clearly more prolific and contrary to this the hatched broods of the type "black spike" are much reduced. Moreover, while selecting their partners, the females show the tendency to prefer males of the type "orange spike". For specimens of both sexes, the authors indicate an average life span of four years.

Total length: Male 8 cm, female 6 cm.

Maintenance in the aquarium and reproduction: The maintenance of these specimens - either in pairs or groups - in soft to slightly hard and slightly acidic water (up to 10°dGH) is problem free. The commented fights among males that occur if the fish are kept in groups (a spacious tank is presupposed) are very much ritualized so that there are almost no serious damages. Kept in groups, in the hierarchic structure, the breeding females take a dominating rank (as it was reported by RÖMER & SOARES, 1996). The process of breeding roughly corresponds to the one given for *Apistogramma cacatuoides*.

Apistogramma linkei KOSLOWSKI, 1985

Meaning of the specific name (etymology): In honour of the author, collector and breeder Horst Linke who besides the type material also caught further specimens for importation and reproduction.

First description: Bonner zool. Beitr. 36 (12), pages 145-162.

Synonyms: None.

Distribution: In the central lowland, the so-called Oriente (= East) of Bolivia in the surroundings of the two cities ofTrinidad and Santa Cruz, there is an area crossed by two tributaries of the Rio Guaporé: the Rio Mamoré and the Rio San Miguel. Between the northern and the southern locality, there is a distance of about 250 km. In the southern part of its distribution range, the specimens can be found sympatrically with *A. luelingi*. In the north, the distribution range of *A. staecki* is attached. The northern and southern races differ a little bit in colouration and size (cp. map page 77).

Habitat: By LINKE & STAECK, the habitats were described analogously as follows: In rest water ponds, ditches and streams which in every case were shadowed by a dense swimming plant layer (*Eichhornia*, *Pistia*). The water measurements were different from those of other areas: total hardness 1.5 to 6°d, carbonate hardness 1 to 6°d, conductivity 37_{29} to 198_{24} as well as a pH from

S03627-4 *Apistogramma linkei*, wild male.

6.6 to 7.8. These variable figures make the fish quite adaptive, therefore they are better qualified for maintenance and breeding than some relatives from purely extreme soft water.

SD and colouration: The intensely yolk yellow breast of the males gave the specimens to their German name ›Gelbbrust-Apistogramma‹. Besides, they show an opulent finnage, whereas the spines of the dorsal fin (the anterior ones excepted) almost have the same length. But the soft rays of the dorsal and anal fin are that much extended that they even reach up to the central part of the caudal fin. On its posterior edge, the caudal fin itself is slightly rounded and its real size can only be seen in threatening behaviour, when it is fully stretched. The throat is whitish, the upper part of the body shows bright olive tones while the tone of the lower part turns over into a tender blue, a colouration that also covers many parts of the fins.

Out of the breeding season, females are beige grey and while courting and breeding they take over only a pale yellow colouration in which the throat and the belly remain white. On the head, a deep black cheek stripe can be seen and instead of a horizontal line a flecked pattern covers the body. The intense caudal blotch is also deep black. The anterior dorsal and anal fin rays then equally take a deep black coloura-tion.

Total length: Male 7 cm, female 4 cm.

Maintenance in the aquarium and reproduction: *Apistogramma linkei* does not place high demands in water quality. For reproduction in an aquarium, the water must have a total hardness of about 4 to 6°d and a pH value around neutral. The more accidential breeding success is now explained by example, because one of the authors (Bork) was able to reproduce this species in the rear part of a photo tank of only 30x20x20 cm. The water readings corresponded to those described before: 5°dGH, pH 7.0 and a water temperature of 26°C. Before the female was put into the rear part of the tank which was separated by a screen, a short plastic tube closed at one end was brought into the gravel ground with an inclination to the background. First the male was kept in a separate tank. When already one week later the female's belly was rounded, the time for a unification of both partners had arrived. For safety reasons, the male was first put into the front part of the tank. But the desired event happened differently because immediately after recognizing the male, the female came to the separating screen and tried to attack the male. The frightened partner stayed at the bottom and showed stress colouration. In this apparently never ending situation and due to the female's swollen belly, only one solution remained: the separating screen had to be removed. But before that, the inner part of the female's cave was controlled: Even without any physical contact, the view of the partner must have stimulated the female to empty its belly and to attach its eggs to the ceiling in the rear part of the tube. Now there was no more reason for the presence of the separating screen. It is hard to believe what can happen in a short period of time of about 10 to 15 minutes during which the author had to leave the room because of an urgent phone call. When he could again concentrate on the scene in the ›photo breeding tank‹, all that had already happened was not supposed to be so soon, nor could it be hoped for: The 80 orange coloured eggs at the ceiling were obviously fertilized, but the male was still (or already again?) lying in the corner, showing stress colouration, as it was true before. But now it could be seen that the male had some injuries in the anal and caudal fin. The female kept on attacking its partner as if she wanted to chase the intruder, but time and again it returned to the entry of the cave and pushed water (and sperm?) into the cave. In order to avoid further injuries, the male was taken out of the tank, although at that time a fertilization of the eggs was uncertain as it could not yet be stated visually. I had to wait. Due to the intense care of the female, it could soon be learned that the eggs developed. After the end of the third day, curiosity could not be suppressed any longer and with a stick the narrowed entry of the plastic tube was freed sufficiently so that the about 80 wriggling larvae could be seen at the ceiling of the tube. At the same time the impression resulted that the hatching rate had been almost 100%. Already at the evening of the eighth day, for the first time, the female lead her young through the aquarium.

Apistogramma luelingi KULLANDER, 1976

Meaning of the specific name (etymology): Dedication name in honour of the collector and ichthyologist Dr Karl Heinz LÜLING. Together with the Swiss A. MEYER, during an ichthyological collecting trip in Bolivia he organized the first specimens and gave them to MEINKEN in preserved state for identification.

Remarks: MEINKEN (1967 and 1969) at first erronuously identified the received material as belonging to *A. borellii*, a name which at that time was still used for *A. cacatuoides*. After

S03636-4 *Apistogramma luelingi*, wild male.

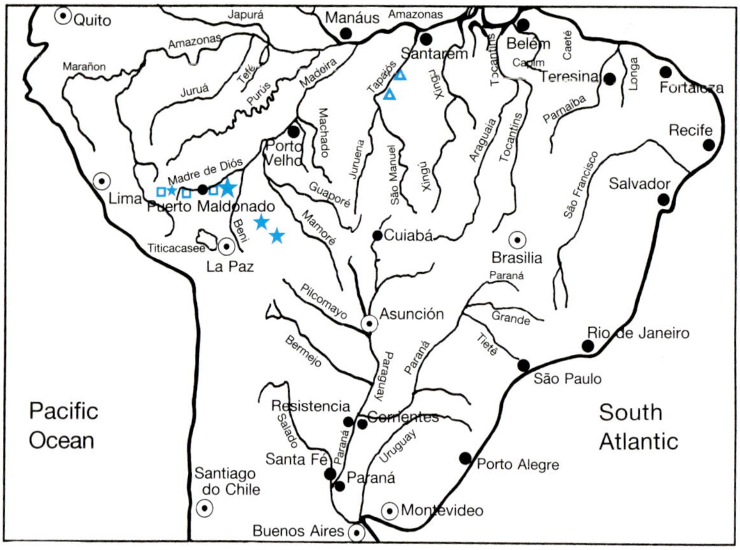

★ *A. luelingi*
□ *A. urteagai*
△ *A. taeniata*

Kullander examined the preserved fish once more and worked on them, he discovered Meinken's error, corrected it and thereupon described the new species.

First description: Bonner zool. Beitr. 27, pages 258-266.

Synonyms: None.

Distribution: In the vast area around Puerto Maldonado/Perú in the drainage of the Rio Madre de Diós as well as to the surroundings of the city of Cuzco (Cuzco Amazónica Lodge). In the Rio Tambopata (Nature Reserve Park Tambopata [Laguna Cocococha] and in the bed of the Rio Manú (National Park of Manú). Linke & Staeck (1992) indicate further localities for this species: Rio Chaparé and Rio Yacapani which belong to the headwater rivers of the Rio Mamoré Therefore these habitats are very far away from the previously mentioned places (cp. map page 86).

Habitat: The sandy bottoms of the partially swampy but also very soft (total and carbonate hardness <1 °d) and very acidic (pH 6.0) waters have aquatic vegetation (*Echinodorus*). Not excluding twigs as well as thick branches and foliage so that the fish which are in constant search for cover can in fact shelter. In zones with a slow current, swimming plants offer additional shelter.

SD and colouration: The males which can reach quite some length have pronounced fins. The membranes of the anterior dorsal fin spines are extended, those of the posterior soft fin rays of the dorsal and the anal are extended with tips and reach up to the central part of the caudal fin. The dark brown bilobated caudal fin is covered with a pattern of bright blue lines and shows longer filaments. The pattern of the caudal fin described here runs into the posterior parts of the dorsal and anal fin. In the upper

Biotope: Rio Madre de Dios,
downstream the city of Puerto Maldonado, still however Bolivan territory
Date: 11.11.1996
Time: 11.00 o'clock
Where? Collections from a side lagoon of the river
Water movement: Very moderate
Water colour: Clear, greenish if seen from above
Surface temperature: 28.6°C

Laboratory results:	
pH-value	6.65
Electrical conductivity (25°C), µS/cm	77
Acid capacity up to pH 8.2 (p-value), mmol/l	n.p.
Acid capacity up to pH 4.3 (m-value), mmol/l	0.70
As carbonate hardness, °d	2.0
Total Alkalis, mmol/l	0.25
As total hardness, °d	1.4
Sodium (Na^+), mg/l	5.6
Potassium (K^+), mg/l	2.3
Calcium (Ca^{2+}), mg/l	6.2
Magnesium (Mg^{2+}), mg/l	2.4
Total Iron ($Fe^{2+/3+}$), mg/l	0.040
Manganese (Mn^{2+}), mg/l	0.005
Ammonium (NH_4^+), mg/l	1.3
Nitrite (NO_2^-), mg/l	0.58
Nitrate (NO_3^-), mg/l	<0.3
Chloride (Cl^-), mg/l	0.3
Sulphate (SO_4^{2-}), mg/l	3.3
Phosphate (PO_4^{3-}), mg/l	0.11
Hydrogen Carbonate (HCO_3^-), mg/l	43
Oxidation	
a)As Potassium Permanganate $KMnO_4$ consume, mg/l	15
b)As Oxygen (O_2), mg/l	3.8

n.p. = not provable

(and occasionally in the lower) border of the caudal fin there is a blue black blotch that is typical for *A. luelingi*! Additionally, the two latter fins are covered by a bluish shimmer. The red line and blotch patterns in the posterior part of the dorsal and caudal fin known from *A. cacatuoides*, can also be seen in some specimens of this species.

Probably besides the extended anterior membranes of the dorsal fin they were decisive for the misidentification of MEINKEN. From the upper lip through the forehead up to the neck and to the anterior dorsal fin rays, runs a golden yellow colour zone that appears to be sprayed on. Contrary to that, the throat and the belly remain white.

In females, the stripe on the cheeks as well as those on the anterior fin rays of the dorsal and the ventrals are deep black. Instead of a black horizontal band, the females of this species show only a striking black lateral blotch.

Total length: Male 8 cm, female 4 cm.

Maintenance in the aquarium and reproduction correspond to what was described for *A. linkei*.

Apistogramma macmasteri KULLANDER, 1979

Meaning of the specific name (etymology): Dedication in honour of the American, Marc MACMASTER who directed KULLANDER's attention to these specimens.

First description: Zool. scripta 8, pages 70-73.

Synonyms: *A. taeniatum* EIGENMANN, 1922 and FOWLER, 1942. These specimens were mentioned by STAECK (1974), SCHMETTKAMP (1976, 1977) and MAYLAND (1978) by the (incorrect) name of their contemporary synonym *A. ornatipinnis*.

Remarks: The species *A. ornatipinnis* created by AHL is a synonym of *A. steindachneri* from Guyana, not of *A. macmasteri*!

Distribution: In the drainage of the upper Rio Meta not far from the city of Villavicencio on the slopes of the eastern Andes (Rio Guatiquia, Rio Ocoa), underneath the Colombian capital (Santa Fé de) Bogotá. A further locality was reported from the Rio Meta (near Puerto López in the province of Meta). Already a few years before the description, the specimens were in trade as a novelty under the trivial name ›Villavicencio-Apistogramma‹ (cp. map page 88).

Habitat: The species can be found in the Colombian Llanos which to the north extend up to Venezuela. There are some streams which also in the arid season do not dry out, even if then their water level is locally only a few centimeters high. Wood, foliage and locally accumulations of aquatic vegetation that appear quite opulent, as well as an additional overhanging vegetation at the banks offer sufficient shelter

The water is extremely soft (total and carbonate hardness <1°d, pH value 6.0 to 5.5, conductivity <30µS). If the water is not shaded by trees, its temperature at noon can pass the 30°C point.

S03652-5 *Apistogramma macmasteri*, wild male, during frontal threatening.

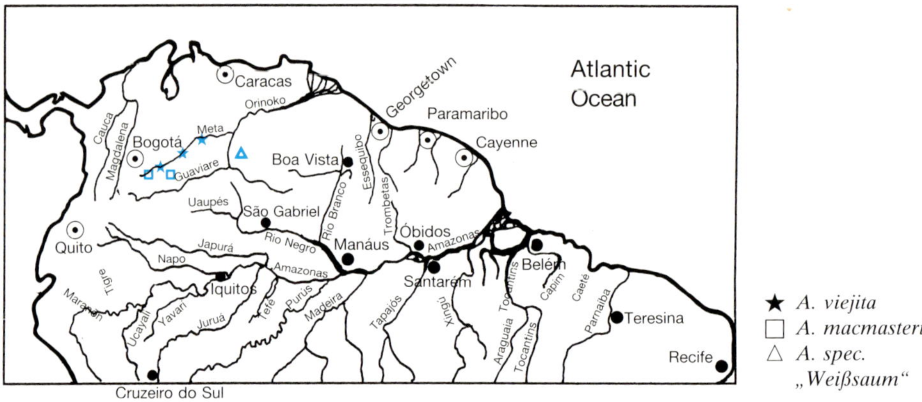

SD and colouration: Fish with red colouration have always had good chances in the hobby to be popular. Indeed, in the wild males of this species (pedigree specimens are also occasionally in the trade!) the red colour traits are mostly limited to the peripheral edges of the caudal fin, a narrow upper edge in the two lobes of the posterior dorsal fin and a few markings on the lower head where they are combined with sky blue patterns.

There are specimens that almost always show their black patterns: dorsal blotches as well as the stripe on the cheeks and the longitudinal band. The latter can either show up continuously in a pattern of black zig-zag lines or as an interrupted line of black rectangles. Similar patterns can also be found in the pale yellow females whose basic colouration is intensified during the breeding season, and whose dark black colouration then is also present on the anterior dorsal fin rays and the ventral fins.

Total length: Males 8 to 10 cm, female 6 cm.

Maintenance in the aquarium and reproduction: For maintenance and reproduction, the length of the aquarium should exceed 80 cm. In order to give the fish a certain security and to take away their inbred shyness, hiding places made of roots and stones should be offered and the rear of the tank should have dense vegetation, covering the ground completely. The aquarium water shall be very soft and acidic. Just the company with other smaller cohabitants gives the Dwarf Cichlids sufficient safety.

As common for *Apistogramma* species, reproduction follows the example of a male-mother-family which means that the female alone cares for the brood, whereas the male takes care of the defence of the territory. Indeed, after the fertilization of the eggs, the male is vigorously driven away from the direct vicinity of the breeding cave, just like any other cohabitant. The male does not have to give its sperm directly to the eggs which can number about 120. Often the female has selected an entry so narrow that the bigger male can hardly pass through it. In this case, the sperm is set free in front of the cave and through fin movements pushed into the cave.

The development from the egg to a swimming young fish happens in the common period of time.

Apistogramma meinkeni KULLANDER, 1980

Meaning of the specific name (etymology): Dedication name to the honour of Dr h.c. Hermann MEINKEN (1896-1976), the many years Director of the Fish Identification Commitee of the VdA who expressed his particular preference for South American Dwarf Cichlids in many works that were also taxonomic.

First description: Bonner zool. Monograph. (14), pages 118-122.

Remarks: The type material comes from a collection realized by the former Belgian King Leopold III. together with the Belgian ichthyologist Dr Jean-Pierre GOSSE in 1967.

Synonyms: None.

Distribution: The specimens live sympatrically (= together) with *A. elizabethae*, *A. brevis* and *A. uaupesi* in the already mentioned habitats near the Rio Uaupés (by the town of Travão) in the northwest of the Brazilian federal state of

S03685-3 *Apistogramma meinkeni*, young wild male.

Amazonas. A further distribution beyond this area into the upper Rio Negro is presently unknown, but it should be probable because - contrary to the other species mentioned - it can be found occasionally in importations from there.

Habitat: The specimens of this species preferably inhabit zones with accumulations of wood and foliage on the edges of sand banks. Water conditions correspond roughly to those in the upper Rio Negro (cp. results on page 59).

SD and colouration: *A. meinkeni* belongs to the *pertensis* group. The specimens equally have a very slender shape as well as a high dorsal fin whose anterior three membranes are free. The big caudal fin is rounded and just like the dorsal fin, has a reddy brown edge. Besides the olive grey greenish iridescent basic colouration, the ventral part stays silvery white. Almost always, even in neutral mood, the black markings (cheek and snout stripes, horizontal band and caudal blotch) are shown. The unpaired as well as the ventral fins have a bluish glimmer and the parts touching the caudal peduncle are covered with rust brown spotted patterns. The ends of the caudal and anal fin are pointed and in specimens that are well kept, in spacious aquaria, with their filaments they reach the centre of the caudal fin. On the cheeks, a pattern of sky blue spots and bent stripes can be seen. The females show a yellowish body colouration. Their dorsal and anal fins are not that much extended.

It should be referred to the fact that a delimitation of this species towards *A. pertensis* can cause headaches for many aquarists. For most authors, there are more or less doubts in one or other direction.

Total length: Male 6 cm, female 4.5 cm.

Maintenance in the aquarium and reproduction: For the reproduction of *A. meinkeni* there is almost nothing known. It has to be assumed that the reproduction is difficult. As the fish comes from areas with pure sandy bottoms, the tank bottom should equally consist of this material.

Apistogramma mendezi Römer, 1994

Meaning of the specific name (etymology): Dedication name in honour of the Brazilian caoutchouc producer, trade unionist and environmentalist Chico Mendes. The author of the book ›Save the Rainforest‹ (1990) which was translated into many languages (among those also in German, 1990) died on 22nd December, 1988, after a murder committed by the son of a Brazilian cattle breeder.

Remarks: The Brazilian spelling of the name ›Mendes‹ ends with an ›-s‹. As indeed due to international regulations only the spelling in the first description is valid, the name ›*mendezi*‹ is the only correct one.

First description: Aqua vol. 1, no. 1 (Journ. Ichthyol. Aquatic Biol., Germany).

Synonyms: none.

S03695-3 *Apistogramma mendezi*, young wild male.

Distribution: To our present (!) knowledge, the specimens live in an area on the right side of the central Rio Negro south of the town of Barcelos which in Brazil became famous for being a reloading point of *Paracheirodon axelrodi*, the Neon Tetra. Further localities are in clear water streams at the upper Rio Negro (cp. map p. 70).

Habitat: The fish occur together with *A. gephyra* and *A. gibbiceps*, mostly in gullies and smaller streams on sandy ground partly covered with stones. The extremely soft clear water has only a weak brownish colouration. Depending on the season of the year, the water conditions change, whereby the pH value mostly remains in a clearly acidic range: 6.0 to 4.2. Electric conductivity was measured between 15 and 80 µS. In the arid time of the year and therefore with the lowest water level, the small Cichlids can be found in depths between 50 and 120 cm above a bottom covered by mud, sand and particularly foliage. Occasionally, aquatic vegetation (*Cabomba* and *Echinodorus*) can be found.

SD and colouration: The species related to the smaller *A. paucisquamis* can be differentiated by the differing pattern of the caudal fin. Whereas the caudal fin of *A. paucisquamis* shows vertical rows of golden spots, the caudal of *A. mendezi* is conspicuous because of marbling bands. In the upper part of the fin, they consist of rows of red dots. The normally high dorsal fin shines in the colours golden yellow, orange and sky blue, the latter one only forming the seam. For it, the sky blue colouration can be found on the lower part of the head, on the anal and ventral fins as well as in the lower part of the caudal fin. In elder males, the ventral fins as well as the tips of the caudal fin can be conspicuously extended. A black horizontal band runs from the upper lip through the eyes and along the

S03695-3 *Apistogramma mendezi*, wild female.

S03695-3 *Apistogramma mendezi*, old wild male.

flanks to the end of the caudal peduncle and even into the base of the caudal fin. Furthermore, in *A. mendezi* three to four stripes on the lower ventral part can be seen. In *A. paucisquamis*, such a stripe can be seen only occasionally.

Total length: Male 7 cm, female 3.5 cm.

Maintenance in the aquarium and reproduction: There are no particular demands in maintenance in an aquarium, they readily consume almost every type of food offered, among these, also deep frozen food. Due to the statements of some aquarists, well adapted specimens with retardation even accept flake food - but this should not be considered a permanent diet!

In order to allow the specimens to reproduce, and to give the eggs best chances to develop, the water has to be adapted as much as possible to the water conditions in the natural habitat of the small Cichlids. This means: very low nitrate concentration, total hardness below 80 µS and a clearly acidic pH value around 5.0. The buffering is not quite easy so that a daily control is indispensable. Under these conditions, the females turned out to be quite productive and in spacious tanks with one male and a ›harem‹ of females a well working reproduction can be started. RÖMER, author of the first description, obviously was one of the first who reproduced this species. Reproduction should not be that easy. Until today, only little is known about the reproduction.

Apistogramma moae KULLANDER, 1980

Meaning of the specific name (etymology): Named after the locality in the area of the Rio Moá in the extreme west of Brazil.

First description: Bonner zool. Monograph. (14), pages 61-65.

Synonyms: None.

Distribution: This representative of the *regani* group has quite probably a rather limited distribution range. The Rio Moá is a tributary of the big Rio Juruá which in its upper reaches in the Brazilian federal state of Acre also drains the border area to Perú. The small and relatively short river ends north of the city of Cruzeiro do Sul. Here there is also the Brazilian Transamazônica No. 364 to the Peruvian Pucallpa (cp. map page 56).

The type material comes from a collection of the former Belgian King Léopold III. and the Belgian ichthyologist Dr Jean-Pierre GOSSE from 1967 (two male specimens from the Igarapé São Salvador).

Habitat: This is a rainforest area where the jungle was heavily cleared in favour of cattle raising. This was also the domain of Chico Mendes (cp. ›Meaning of the specific name‹ in *Apistogramma mendezi*). On page 91, the results of the latest examinations in the drainage of the Rio Moá are given. Despite the clearing activities, the typical tropical water qualities can still be found.

S03715-3
Apistogramma moae, wild male.

SD and colouration: Other than the normal metallic and shimmering general impression of the Dwarf Cichlids, this species is without attractive colouration, unfortunately making them not the most popular representatives of their genus. Specimens of both sexes have a rounded caudal fin. Bigger males can be easily detected by the pointed tips of the caudal and anal fin. In the females, these are rounded. In specimens of both sexes, a discontinued narrow horizontal band is present, ending in front of the base of the caudal fin in a vertically oval blotch.

The female is coloured brightly yellow brownish. Belly and breast stay whitish. The ventrals are colourless and their tips are yellowish. The mouth and the throat are bright

Flooded lagoon in the drainage of the Rio Moá. Habitat of *A. moae*.

Biotope: Rio Moá,
approx. 25 km upstream of Cruzeiro do Sul and approx. 20 km upstream the mouth of the Rio Juruá
Date: 15.11.1996
Time: 10.00 o'clock
Water movement: In the main river it was strongly current
Water colour: Clear and lightly tea coloured
Surface temperature: 26.2°C

Laboratory results

pH-value	6.60
Electrical conductivity (25°C), µS/cm	28
Acid capacity up to pH 8.2 (p-value), mmol/l	n.p.
Acid capacity up to pH 4.3 (m-value), mmol/l	0.28
As carbonate hardness, °d	0.78
Total Alkalis, mmol/l	0.10
As total hardness, °d	0.28
Sodium (Na^+), mg/l	1.4
Potassium (K^+), mg/l	1.2
Calcium (Ca^{2+}), mg/l	2.9
Magnesium (Mg^{2+}), mg/l	0.66
Total Iron ($Fe^{2+/3+}$), mg/l	1.0
Manganese (Mn^{2+}), mg/l	0.006
Ammonium (NH_4^-), mg/l	0.02
Nitrite (NO_2^-), mg/l	<0.02
Nitrate (NO_3^-), mg/l	0.5
Chloride (Cl^-), mg/l	0.2
Sulphate (SO_4^{2-}), mg/l	0.9
Phosphate (PO_4^{3-}), mg/l	0.11
Hydrogen Carbonate (HCO_3^-), mg/l	14.6
Oxidation	
a) As $KMnO_4$ consume, mg/l	31
b) As Oxygen (O_2), mg/l	7.9

n.p. = not provable

brown. Below the eyes, there is a pattern of bright blue stripes.

The basic colouration of the males is olive brown, the entire body and all the fins are covered by a blue glimmer. In a state of affection, the blue can be very beautiful and then covers the entire body including the fins. The cheek stripe is only slightly pronounced.

From 1987 to 1988, *A. moae* was occasionally offered in trade. At present, this species can be seen very rarely in the hobby.

Total length: Male 7 cm, female 4 cm.

Maintenance in the aquarium and reproduction: The reproduction of this species in the aquarium has succeeded several times and seems to be similar to the closely related *A. cruzi*.

Apistogramma nijsseni Kullander, 1979

Meaning of the specific name (etymology): Dedication name in honour of the Dutch ichthyologist and Curator of the Zoological Museum Amsterdam, Dr Han Nijssen, who gave the type specimens caught and preserved by the Swiss Patrik de Rham to the author.

First description: Rev. suisse Zool. 86 (4), pages 937-945.

Synonyms: None.

Remarks: The eight specimens collected by de Rham in October, 1977 in Perú were without exception females. As at the time of the description about form and colouration of male specimens nothing was known, another collecting trip by the same collector in August/ September 1981 could bring clarity. It was expressed in a detailed article (unfortunately only in French with a short summary in English) which de Rham & Kullander published in 1983. Therein, besides others the authors propose the trivial name ›Panda Dwarf Cichlid‹ which is easy to maintain.

Distribution: In the beginning it was assumed that the species was spread exclusively in a relatively limited area in the bed of the Rio Ucayali (province Loreto/Perú) along the road/course from Jenaro Herrera (Rio Ucayali) to Colonia Angamos (Rio Yavarí - Quebrada Copal, Quebrada Carahuayte), about 14 km from the first mentioned town. There, these specimens occur

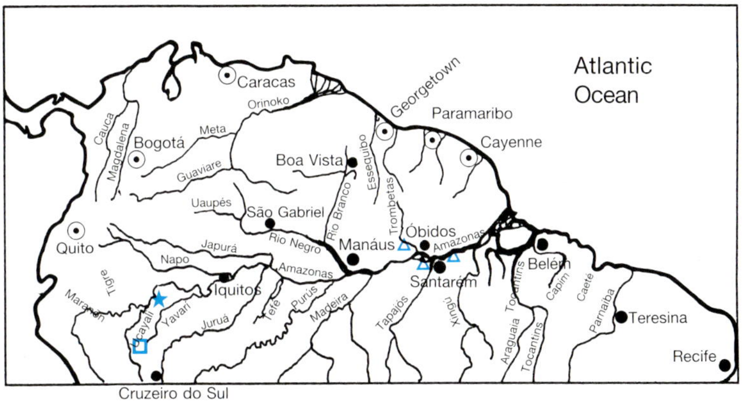

★ *A. nijsseni*
□ *A. spec. „pandurini"*
△ *A. spec. „Smaragd"*

S03725-4 *Apistogramma nijsseni*, wild female without lateral blotch, threatening.

together with *A. agassizii* (red form), *Aequidens/ Bujurquina* species as well as with representatives of the genus *Laetacara* (*L. fluviatilis*).

Due to newer imports, it has meanwhile to be supposed that *A. nijsseni* inhabits a distribution range that is much bigger (cp. map page 93). Obviously, its distribution also overlaps with *A. cacatuoides* as well as some of the still undescribed species that were introduced recently.

Habitat: The clear rust to dark brown water is rich in humine acids and therefore the values measured in consecutive years (1981/1983) by DE RHAM as well as by LINKE & STAECK are as follows: water temperature 25 to 27.5°C, total and carbonate hardness <1°d, conductivity 3 to 27.5 µS, pH value 5.4 to 5.6 and oxygen 6.45 ppm. The data in KUSTERS (1991) is similar: 15°S, pH value 5.5, total and carbonate hardness <1°d and a water temperature of 25°C.

The habitats that are knee deep to about 1 m, where the water has a slight current, are in the area of a former rainforest whose existence is threatened by many clearages. The sandy bottoms are covered with lots of foliage, branches and twigs. Crayfish in various phases can be found (mostly belonging to the genus *Macrobrachium* of which alone in Amazonia there are about 35 different species that are difficult to identify). As in black water mosquito larvae do not develop easily, small and average sized fish preferably feed from crayfish.

SD and colouration: *A. nijsseni* belongs to the most beautiful Dwarf Cichlids, a statement not only referring to the males, but also to the females. Specimens that have such patterns of big black blotches on the gill covers and on the centre of the flanks, cannot be found in any other South American Dwarf Cichlid. The basic yellow body colouration of the females is intensified during the courting and breeding period. The equally intense black colouration forms the markings of the anterior dorsal and anal fin rays as well as the caudal blotch, these are all visible in the common colour pattern. During the courting procedure, the black blotches can occasionally turn over into green enamel-like iridescent tones.

The big hefty males show yellowish to brass yellow tones on the throat, breast and belly. The

S03725-4 *Apistogramma nijsseni*, wild female with lateral blotch.

S03725-4 *Apistogramma nijsseni*, wild male.

posterior part of the body can be covered by a blue glimmer. The dorsal fin has a straight edge. The second to the fourth membrane show the beginnings of a slight extension. Contrary to that, the anal fin rather, shows the yellowish colouration of the ventral and breast part. This is similar for the ventral fins. The caudal fin which is rounded at its posterior edge means the colour attraction of this species: The caudal blotch is followed by a pale yellow zone in which the fin rays shimmer iridescently. The marginal seam is intensely orange red; then there is a sooty black submarginal band.

In offspring specimens the colours glimmer less intensely and the extended parts of the fins are searched for in vain.

There are various colour forms. Some local varieties are no longer yellow in both sexes; a dirty grey brown is rather dominant. But hobbyists considered these varieties less attractive. It should also be referred to the occasional critics of breeders: The blotches in the females are said to be no longer ›beautiful‹, due to unsatisfactory selection or faults while breeding. Also in this concern, the variability of the breeding specimens can be of decisive importance!

Total length: Male 9 cm, female 6 cm.

Maintenance in the aquarium and reproduction: The representatives of this species are not very demanding when it comes to water quality in the aquarium - as long as reproduction is not concerned. For breeding, one should offer the water conditions mentioned before (do not forget the low nitrate concentration!). The reproduction of these small Cichlids also succeeds in water of average hardness, but then it is clearly less productive! For an optimal breeding, a total hardness of 50 µS turned out to be ideal. Then, with 60 to 70 offspring, it can be quite productive. An additional filtration through peat is to be recommended. In order to keep the pH around 5.5 in a controlled way, the addition of oak extract is recommended. As very soft water is a very bad buffer, it has to measured and - if necessary - dosed repeatedly. Besides this, of high importance is a spacious tank of at least 80 cm (100 cm are better) in which also aquatic plants can grow (as long as they still grow under the conditions described here!), Females that are fed well with diversified food are one presupposition for an acceptably productive reproduction. Once more, the previously described behaviour of a male-mother-family is in the foreground, whereby the female alone has to care directly for thc fry, whereas its partner takes care of the safety of the territory. As an average, 60 to 70 reddish to orange red eggs are attached to the ceiling of the cave at their poles. While the female is looking after the brood, it often shows an admired golden yellow colouration upon which the enormously enlarged deep black throat, lateral and caudal blotches are shining.

At a water temperature of 27°C, after almost four days the larvae leave the eggs and after five to six more days they develop into young fish that are able to swim. Once the larvae are hatched, the partial water change and the longing for a low nitrate concentration in the breeding tank are important and the former should not be put off. But the water readings must not be changed. Controlling the water and - if necessary - constant regulations are important to the young fishes development. When well fed (in the beginning with *Artemia* nauplii), the young fish grow fast.

The mathematic inequality of the sexes which was mentioned in the beginning of this book, whereby in the most characteristic case only male or female offspring are obtained (monogeny), was frequently reported for specimens of this species. In serial experiments under controlled conditions, RÖMER (1989) found out that the number of male offspring is higher with raising temperature! On the other hand, in a temperature range between 20 and 23°C there are no males among the young. Between 23.1 and 25.0°C the rate of males reached about 18%, whereas at a temperature between 25.1 and 27.0°C the sex ratio with 57% males was almost counterbalanced. If the breeding temperature was raised to 27.1 to 29°C, 82% males were obtained. A ›top male result‹ was obtained at a breeding temperature between 29.1 and 32.0°C: 100% of the offspring belonged to the male sex. In this concern, an additional reduction of the pH value between 6.9 and 4.5 has to be considered. But it affected the sex ratio in a less differentiated way than the temperature of the breeding water. »The water hardness could not prove an affect on the sex ratio, but it showed a clear influence on the development rate of the eggs. This rate was raised from average 16% to 83% by a lowering of the total hardness from 10 to less than 0.5°d«, noticed the mentioned author.

Apistogramma norberti STAECK, 1991

Meaning of the specific name (etymology): Dedication name in honour of the discoverer Norbert WISHEU who discovered the specimens during a collecting trip in the Peruvian province Loreto in May/June 1989 respectively 1990. The specimens were at first called by the trivial name *Apistogramma* spec. ›Großmaul‹.

First description: Ichthyol. Explor. Freshwaters 2 (2), pages 139-149.

Synonyms: None.

Distribution: The type locality of this species which is placed into the *cacatuoides* group is a left sided jungle stream covered by trees which flows into the Quebrada Nuevo Horizonte (in the drainage of the lower Rio Tahuayo, about 73°05'W, 04°05'S). The area is upstream of the Amazon near the city of Iquitos.

After the first imports, also commercial exports repeatedly brought in specimens of *A. norberti*. Due to this, the distribution range of the species seems to be wider than formerly supposed (cp. map page 51). For example, in import shipments with *A. nijsseni* repeatedly specimens of *A. norberti* were found. The colouration of the males can vary from intensely pronounced colours to a more slightly grey green colouration. Particularly the zone around the mouth and the third to fifth dorsal fin ray can be more or less intensely orange red.

Habitat: Following information from the author (STAECK) who himself brought further type material to Germany, the habitat is a »very small forest stream, one to two meters wide, shaded, with a considerable descent, draining the rainforest which there is markedly hilly. In early August, in the time of low water level, only the deeper parts of the streambed had water. These accumulations of rest water which were not connected with each other, covered only a surface of one square meter and had a maximum depth between 30 and 60 cm.«

The water readings in the habitat of the specimens were measured at 24°C with 10 µS as well

S03742-4 *Apistogramma norberti*, wild male during lateral threatening.

as a total and carbonate hardness below 1 °d and a pH value of 5.9. Besides twigs and thicker branches, the bottoms of these narrow streams show a strong accumulation of foliage all over. Once more, among the accompanying fish fauna, the presence of a predatory Characin (this time it was *Erythrinus erythrinus*) was discovered. These fish can be found in most *Apistogramma* habitats. In his work, the author refers to the fact that the specimens of *A. agassizii* and *A. eunotus* that were found a few hundred meters away did not occur sympatrically with *A. norberti* and comes to the conclusion »that *A. norberti* has a particular ecological niche and different necessaries of life« than the other two representatives of the genus.

SD and colouration: The previously mentioned trivial name can be attributed to the fact that the males have a mouth that is conspicuously big for this genus which can even surpass that of *A. cacatuoides* and first of all is dominated by a mighty slightly protrusive lower jaw.

Specimens from different areas can show a more or less intense colouration (cp. ›Distribution‹). Specimens of both sexes show a similar body colouration and pattern which indeed shows that dominant males tend to suppress the lateral band and the lateral blotch. In specimens of both sexes the two (or three) anterior dorsal fin rays are deep black and shorter than the three (or two) extended following ones. In the long tips of the membranes, these rays are equally coloured orange as are the lips and the part below the eyes of male specimens.

The dorsal fin has a fine sooty black seam. As a trait which until today is unique in the *cacatuoides* complex, a black blotch can be seen, which in male specimens is situated in the basal part of the soft fin rays in the dorsal fin, but in females it can not be obviously found. In both sexes, the caudal fin is rounded, and dominant males show a black seam. All unpaired fins and the ventral fins are iridescent bright blue. The central part of the caudal fin is covered with rows of bright blue dots.

Similar to specimens of *A. nijsseni*, also the upper and posterior flanks are covered by a blue iridescent glimmer. Occasionally, males also show two rows of stripes on the lower part of the body or near the abdomen.

S03742-4 *Apistogramma norberti*, wild male.

Females are beige and show an intense cheek stripe as well as a lateral band with a lateral blotch. Starting in the gill area, the whole body is covered with a tender green iridescent tone.

Total length: Male 8 cm, female 6 cm.

Maintenance in the aquarium and reproduction: Reproduction was already achieved in quite some extent and is similar to that for *A. nijsseni*. In the breeding colouration, females show a shining yellow and the mentioned tender green iridescent tone can not be seen any more. But now the lateral blotch as well as the cheek stripe stand out a very concise deep black.

Apistogramma ortmanni (EIGENMANN, 1912)

Meaning of the specific name (etymology): Dedication name in honour of the collector Ortmann. Unfortunately, the author (EIGENMANN) does not give any information about further coherences.

First description: Mem. Carnegie Mus. 5, pages 506-507.

Synonyms: *Heterogramma ortmanni*.

Distribution: In Guyana (in the systems of the Essequibo River and its tributaries like the Potaro River, Rupununi River [Gluck Island, Konawaruk and Rockstone in the Essequibo system; Kangaruma, Kuribong, Amatuk, Erukin in the Potaro system and "Packeoo Fall" and Karanambo in the Rupunumi system]); in Surinam (Corantijn River, Lucie River [Kabalebo River downstream Avanavero Vallen near the mouth of the Lucie River; Nickerie District, between the Dalbana Creek and Amotopo]) as well as in Venezuela (Rio la Clarita/Estado Bolívar, Rio Cuyuni) (cp. map page 141).

Habitat: In the distribution ranges mentioned for Guyana and Surinam, even today mainly rainforest dominates the landscape. But in the Venezuelan Estado Bolívar with the Rio Cuyuni

Biotope:
Venezuela, Rio Apónguao,
a tributary of the Rio Caroni, whose headwaters - seperated by a water shed - are in the same area as the Rio Cuyuni
Date: 11.2.1991
Time: 10.00 o'clock
Water movement: Rapidly flowing
Water depth: Very different, not established in detail
Water colour: Tea coloured

Laboratory results:	
pH-value	6.39
Electrical conductivity (25°C), µS/cm	9
Acid capacity up tp pH 8.2 (p-value), mmol/l	<0.05
Acid capacity up to pH 4.3 (m-value), mmol/l	0.09
As carbonate hardness, °d	0.25
Total Alkalis (Ca^{2+} + Mg^{2+}), mmol/l	n.p.
As total hardness, °d	n.p.
Sodium (Na^{+}), mg/l	0.8
Potassium (K^{+}), mg/l	0.6
Calcium (Ca^{2+}), mg/l	<0.1
Magesium (Mg^{2+}), mg/l	<0.1
Total Iron ($Fe^{2+/3+}$), mg/l	0.19
Manganese (Mn^{2+}), mg/l	0.002
Ammonium (NH_4^{+}), mg/l	0.03
Nitrite (NO_2^{-}), mg/l	<0.02
Nitrate (NO_3^{-}), mg/l	<0.1
Chloride (Cl^{-}), mg/l	0.25
Sulphate (SO_4^{2-}), mg/l	<0.1
Total Phosphate (PO_4^{3-}), mg/l	<0.04
Hydrogen Carbonate (HCO_3^{-}), mg/l	5
Oxidation	
a) As Potassium Permanganate consume $KMnO_4$, mg/l	18
b) As Oxygen (O_2), mg/l	4.6

n.p. = not provable

S03755-3 *Apistogramma ortmanni*, wild male.

and in the ›Zona en reclamación‹ (a vast land claim of Venezuela to Guyana) the rainforest has already vanished.

Indeed, all over the vast distribution range the water is not uniform, but it is very soft with a clearly acidic pH value. Even in these areas, the habitats with shallow water should be similar to those in the areas of the other *Apistogramma* species. In the places preferred by the small Cichlids, the bottom of the streams and gullies which here is also sandy is covered by twigs, coarser branches and foliage. The tea coloured water allows only limited growth of aquatic plants.

SD and colouration: *A. ortmanni* is a relative of *A. regani* and therefore it is to be placed in the *regani* group. The males show a quite irregular lateral band which ends in front of the base of the caudal fin. At its end, there is vertical oval caudal blotch. Up into the horizontal band, the neck and the dorsal part are olive yellow. The ventral part is bright yellow to whitish. Compared to other *Apistogramma* species of the group, the males of *A. ortmanni* show a relatively big (average high to high) dorsal fin. It is also olive yellow and the membranes - particularly in the anterior part - are sooty black. In the caudal fin, one can see vertical rows of brown red spots. Between these, the caudal fin glimmers bright blue. The anal fin has bright blue stripes. On its posterior part, one also finds brown red dots. In state of affection, the colouration generally becomes darker. Behind the front third of the body are five vertical stripes that are dark olive beige.

Total length: Male 9 cm, female 6 cm.

Maintenance in the aquarium and reproduction: *A. ortmanni* is not very pretentious. The specimens turned out to be relatively adaptable, but the males show - similar to most of their relatives in the genus - a strong intraspecific aggression. During breeding attempts, the water conditions should be as close the natural value as possible. Water poor in minerals with slightly acidic quality is necessary for the development of the fry. Due to their not very attractive colouration, and because of the few imports from Guyana (the former British Guyana), in the recent years the maintenance of these specimens regressed. In 1983, WACHTER describes the reproduction in a 50 liter tank, whereby he states that the courtship of the male can turn into aggression only if the female is withdrawing to escape. Hereby, he comes to the conclusion that besides the considerable magnification of the habitat (in the aquarium), a competitor should be present in the territory and perhaps also the relationship female : male (here 1 : 1) should be changed to the favour of the females.

Apistogramma parva AHL, 1931

Meaning of the specific name (etymology): From the Latin adjective ›*parvus*‹ = small. The feminine form with the ending "a" is adapted to the generic name.

First description: Sitzber. naturforsch. Freunde Berlin, pages 210-211.

Synonyms: None.

Remarks: Due to today's requirements, the type material, a young fish of 15.8 mm standard length and 22 mm total length, can not alone give satisfactory determination readings any longer. Therefore, KULLANDER (1980) stated that the specific status has to be considered ›uncertain‹ until new collections from the river are available.

Examinations of other scientists are said to have shown (RÖMER pers. comm. 1996), that it is a young specimen that has to be placed in *A. agassizii*, after which a synonymization with this species would be justified.

Apistogramma paucisquamis KULLANDER & STAECK, 1988

Meaning of the specific name (etymology): A term derived from Latin which means ›sparsely scaled‹. In the name definition, the author (KULLANDER) refers to the relatively low number (12) of the scale rows around the caudal peduncle. As representatives of other species of this genus show 16 and more rows (for example, 12 arc found in Taniacare), concerning this, specimens of this species are of special quality.

First description: Cybium 12 (3), pp 189-201.

Synonyms: None.

Distribution: In the drainage of the central and lower Rio Negro where the fish stay in shallow water near the banks. During collecting trips to the Rio Negro and hereby especially in the area of the Anavilhanas Island group, the small Cichlides were in my nets several times (cp. map page 68).

Habitat: *A. paucisquamis* can be mostly found in shallow water, near the banks. These close relatives of *A. mendezi* live in the dense foliage covering the bottom consisting of sand and sediments. As the river is very poor in nutrients and therefore has almost no insect larvae (the absence of mosquito larvae makes the stay in uninhabited areas a pleasure!), but as all over the year freshwater crayfish are present in many species and stages of life, the Cichlids living here feed mainly from them.

SD and colouration: Before its scientific description, this small and slender species was given the trivial name ›Glanzbinden-Apistogramma‹ by SCHMETTKAMP (1981) and already at the Amazon locality, the males occur with several colour forms. In colour, discriminating traits can be seen in the ventral part, in the caudal fin as well as in the tips of the ventral fins. The latter can show a red orange, yellowish, grey or pure white tone with a blue shimmer. A variable colour pattern of the ventral part is developed. In adult specimens, it can be golden yellow. Another difference can be seen in the bilobated

Biotope: Anavilhanas,
Island archipelago in the lower part of the Rio Negro (nameless lagoon)
Date: 24.9.1986
Time: 18.10 o'clock
Water movement: Very slowly flowing
Water depth: 5-6m
Water colour: Tea brown

Laboratory results:

pH-value	5.82
Electrical conductivity (25°C), µS/cm	11
Acid capacity up to pH 8.2 (p-value), mmol/l	<0.05
Acid capacity up to pH 4.3 (m-value), mmol/l	0.10
As carbonate hardness, °d	0.3
Total Alkalis (Ca^{2+} + Mg^{2+}), mmol/l	0.02
As total hardness, °d	0.1
Calcium (Ca^{2+}), mg/l	0.46
Magnesium (Mg^{2+}), mg/l	0.20
Total Iron ($Fe^{2+/3+}$), mg/l	0.22
Manganese (Mn^{2+}), mg/l	0.02
Ammonium (NH_4^+), mg/l	0.06
Nitrite (NO_2^-), mg/l	<0.02
Nitrate (NO_3^-), mg/l	<1
Chloride (Cl^-), mg/l	3
Sulphate (SO_4^{2-}), mg/l	8
Total Phosphate (PO_4^{3-}), mg/l	<0.04
Hydrogen Carbonate (HCO_3^-), mg/l	6
Oxidation Potassium permanganate $KMnO_4$ consume, mg/l	319

S03770-3 *Apistogramma paucisquamis*, adult wild male, ›Orange seam‹.

caudal fin that has white dots. In some specimens, this fin has an upper and a lower orange seam, in others it is absent. Of the three pattern varieties mentioned, the grey blue is rarer. In this form, the golden and orange tones are missing completely. The colour form with the golden yellow belly is more frequent. In the males, not only the ventral part shows an intense golden yellow colouration; the throat shows either an orange red or golden yellow area.

By many aquarists, the form considered the most beautiful is when cheeks, throat and ventral part show a clearly orange tone and additionally the upper and lower edge of the caudal fin are clearly coloured orange red. In adult males, the big caudal fin is bilobated and shows a pattern

S03770-3 *Apistogramma paucisquamis*, juvenile wild male, ›Orange seam‹.

S03770-3 *Apistogramma paucisquamis*, old wild male, gold breast variant.

of white to rust red dots. The fish owe their German name to the shining scale rows above and below the clearly black lateral band. On the central part of the body, one sees a lateral blotch. Particularly by the females which during courting and breeding time is mostly brass yellow, and two blotches can be shown.

Total length: Male 5 cm, female 3.5 cm.

Maintenance in the aquarium and reproduction: For maintenance and breeding of *A. paucisquamis* one should care for a sufficient number of hiding places. Despite their size, these small Dwarf Cichlids can become very aggressive towards each other. Females that are not ready to spawn have to be able to evade the attacks of the very actively courting males, otherwise under certain circumstances serious injuries can occur. Therefore, the reproduction of this little cave breeder in the aquarium is not at all easy. The specimens require a preferably soft water that is poor in nitrate and has a pH value between 5.6 and 6.0. The egg laying of the small females comprises of (only) 40 to 45 reddish eggs which do not stick together but are attached to the cave ceiling with one pole with smaller clearances. At a water temperature of 27°C, the larvae develop in four days up to hatching and after four to five more days, the young swim free.

S03770-3 *Apistogramma paucisquamis*, adult male, gold breast variant.

S03770-3 *Apistogramma paucisquamis*, juvenile wild female.

Apistogramma payaminonis KULLANDER, 1986

Meaning of the specific name (etymology): From the locality in the system of the Rio Payamino, a left sided tributary of the upper Rio Napo in the Ecuadorian province of Napo.

First description: Cichlid fishes of the Amazon River drainage of Peru. Swed. Mus. Nat. Hist., Stockholm (431 pages), pages 184-188.

Synonyms: None.

Distribution: Hitherto found only in the system of the Rio Payamino (Quebrada Ahuano upstream of Ahuanopaccha, at the upper Rio Tutapishcu, some kilometers south-west of the town of San José de Payamino) as well as in a forest stream near the Rio Napo upstream the mouth of the Payamino close to the town of Puerto Coca, the former Francisco de Orellana (cp. map pages 10/11 above left and map page 181).

Habitat: KULLANDER mentions the information given by D.J. STEWART who made available a part of the type material, collected in November 1983: water temperature 22.5°C, pH value 6.4, width of the river 8 to 10 meters. The specimens were captured near the banks of the damed water from where they wanted to escape quickly to the centre of the river. Habitat depth 50 to 100 cm. The bottom here consists of a naturally grown stone, in the damed part covered with coarse branches and sand.

SD and colouration: Together with the closely related *A. nijsseni*, *A. payaminonis* is closely associated with the *cacatuoides* group. In particular, the posterior marking of the caudal fin of the males reminds us of the family of *A. nijsseni*. But in his work of 1986, KULLANDER refers to the fact that for a more detailed hypothesis about the family relationships further studies are necessary. So far not yet live introduced.

Total length: The biggest specimen of the type material (holotype male) is given with a standard length (= measured without the caudal fin) of 39.6 mm, which should correspond to a total length of 6 cm.

Maintenance in the aquarium and reproduction: For the time being the necessary information is missing (see above).

Apistogramma personata KULLANDER, 1980

Meaning of the specific name (etymology): Adjective form of the Latin ›*personatus*‹ = ›masked‹, ›provided with a mask‹. The ending "a" corresponds to the rules of adaption to the female gender of the generic name.

First description: Bonner zool. Monograph (14), pages 111-114.

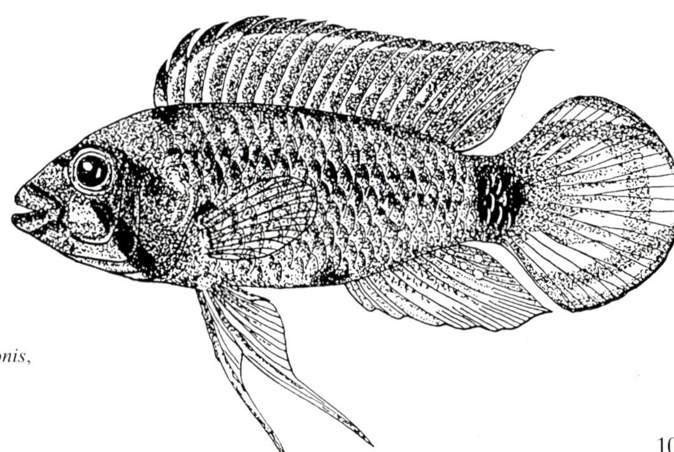

S03774-4 *Apistogramma payaminonis*, appearance of any adult male.

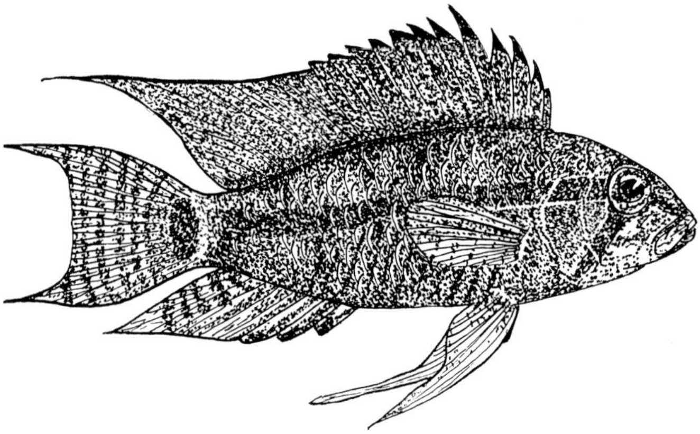

S03780-3 *Apistogramma personata*, appearance of any adult male.

Remark: The description of this species was done on specimens which the former Belgian King Léopold III. collected together with the Belgian ichthyologist J.P. GOSSE in 1967.

Synonyms: None.

Distribution: As the only hitherto known type locality, the Rio Uaupés near Assaí in the northwest of Brazil, not far from the Colombian border (where the river is called Vaupés) (cp. page 68).

Habitat: The specimens live in shallow water near the banks above accumulations of foliage. The water is extremely poor in minerals (conductivity 10 µS) and the pH value of about 4.8 is clearly acidic.

SD and colouration: Unknown.

Total length: In his first description, the author KULLANDER reports a total length of 65.3 mm for the biggest male (holotype). The standard length (SL = without caudal fin) is noted as 49.2 mm. For the biggest female, 44.2 mm total length are given.

Maintenance in the aquarium and reproduction: As hitherto the specimens could not yet be imported alive, no corresponding information can be given.

Apistogramma pertensis (HASEMAN, 1911)

Meaning of the specific name (etymology): Feminine form of the adjective of the Latin verb ›*pertendare*‹ = ›to assert oneself‹, ›to persist in‹. The author did not give any information about how he wanted the term to be interpreted. Therefore, SCHMETTKAMP (1992) created the hypothesis the author (HASEMAN) would allude to the fact »that the species orignially described as variety of *A. taeniata* would be able to "assert itself" (which indeed is true) or wanted to refer to the close relationship to *A. taeniata* by "asserting itself"«.

First description: Ann. Carnegie Mus. 7 (3-4), page 359.

Remarks: HASEMAN described the species as variety of *A. taeniata* (which can be understood comparatively to a subspecies in the sense of modern taxonomy). Already REGAN (1912) and DE MIRANDA RIBEIRO (1918) raised *A. pertensis* to the rank of a valid species. FOWLER (1954) on the other hand listed *A. pertensis* only as a subspecies of *A. taeniata*. Among themselves, the species *A. iniridae*, *A. gephyra*, *A. pulchra*, *A. meinkeni* and *A. uaupesi* which were described in the recent decades, show clear family relationships. MEINKEN's opinion (1961) saying that *A. pertensis* also could exist upstream of Leticia in Perú, is considered more than doubtful by KULLANDER.

Synonyms: *Heterogramma taeniatum pertense*.

Distribution: The proven distribution range reaches from Santarém in the Brazilian federal state of Pará (right side of the Amazon) through the mouth of the Rio Negro up to Manacapurú (left/Solimões side of the Amazon, together with *A.gephyra*), into the Rio Negro (Anavilhanas Island group, together with *A. regani*). Furthermore, representatives of the species were found in streams near the Rio Preto da Eva (which flows into the Amazon east of Manáus) and in the Igarapé Mapiri (together with *A. agassizii*) (cp. map page 79).

S03787-4 *Apistogramma pertensis*, adult wild male.

Habitat: In slightly differing habitats where the water is soft throughout, but because of its mixing ratio black/white or black/clear it can have a different pH (acidic in any case!). As in the areas mentioned here rainforest is dominant, one can assume that the sandy bottoms near the banks are covered with twigs, coarse branches and big accumulations of foliage.

SD and colouration: The species which gave its name to the pertensis group belongs to the slender Dwarf Cichlids which do not shy away from an existance in water with a slightly strong current. Depending on their origin, the specimens can differ from each other in body and fin shape. In some specimens, the big eye which allows an existance in relatively dark water, is con-

S03787-4 *Apistogramma pertensis*, wild female.

spicuous. Males have a hefty head, equally formed lips and the corners of the mouth are directed downwards. The height of the dorsal fin is average to sail-like. In male specimens, the tips of the dorsal and anal fins are pointed and extended, in the females they are rather rounded. The caudal of the males is also rounded. Additionally, it is covered by a pattern of bent rust red bands of dots. The cheek and the horizontal band are mostly well recognizable, the same is true for the caudal blotch at the base of the caudal fin.

Here it should be referred to the difficult demarcation of the species against *A. meinkeni*. Finally, concerning this most authors have again and again slight doubts.

Total length: Male 7 cm, female 5 cm.

Maintenance in the aquarium and reproduction: *A. pertensis* is kept only rarely. Indeed, in the recent years the species was offered in the trade more frequently. Already for maintenance, the slender elegant specimens need soft, acidic water that is poor of nitrate. Only under these conditions, can the specimens show their modest beauty that has a unique attractiveness. If kept under worse conditions, the specimens are merely silvery grey.

About reproduction nothing is known. With some certainty, for this reason breeding this species can be considered one of the most difficult tasks of an aquarist.

Apistogramma piauiensis KULLANDER, 1980

Meaning of the specific name (etymology): The elongated Brazilian federal state of Piauí with its capital Teresina and the second largest city of Parnaíba (upstream the mouth of the river of the same name) is in the north-east of the big country. The specific name was derived from the name of the state.

First description: Bonner zool. Monograph. (14), pages 79-83.

Synonyms: None.

Distribution: Only known from the area around the type locality in the Estado do Piauí: Lagoa Seca and Rio Paranaíba near Barra do Longa, close to Buriti dos Lopes at the mouth of the Rio Longa. The Rio Parnaíba rises in the southern situated Serra do Pauí.

Further collections in the greater surroundings of this area should follow (imports have already been carried out) in order to discover related species hitherto unknown in the area between the Rio Tocantins and the Rio Panaíba or to solidify the coherences hitherto considered (cp. map left page).

With this distribution range, *A. piauiensis* is the hitherto eastern most occuring representative of its genus. It can well be placed into the regani group and its closest relatives are *A. caetei* and *A.* spec. ›Paraguay I‹.

Habitat: Following the depictions of the US-American collector and ichthyologist T.R.

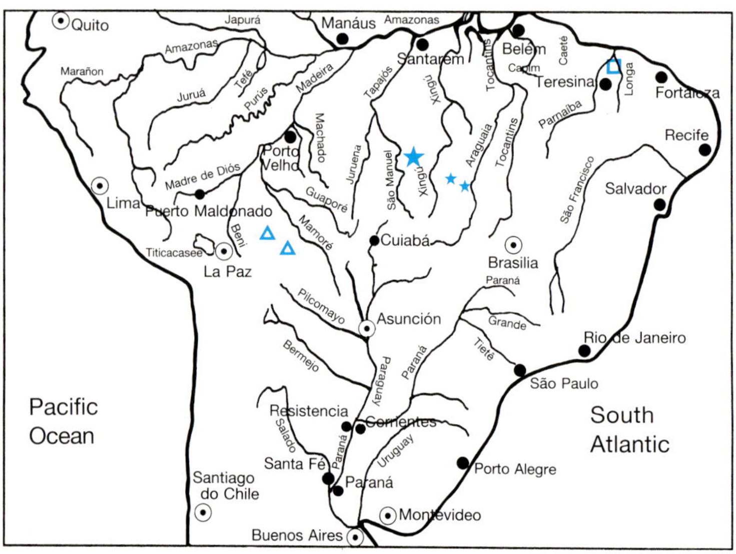

★ *A. spec. „Peixoto“*
□ *A. piauiensis*
△ *A. staecki*

S03795-4 *Apistogramma piauiensis*, adult male above dark ground.

ROBERTS: The Lagoa Seca (›= dry lake‹) is situated one or two kilometers from the main river of the Rio Parnaíba and the town of Barra do Longa (›long bar‹) and belongs to the inundation zone of the river. At the time of the collection, the area was flooded up to one meter over a zone that was about 200 to 400 meters wide and one kilometer long. The ground was muddy and interspersed with some rocks. There was no remarkable higher vegetation. Water readings at the moment of the collection were not mentioned by the mentioned ichthyologist.

<u>SD and colouration:</u> It was only in spring 1995 and 1996 when a few imports came in.

As its closest relatives, *A. piauiensis* also does not show particular traits in the finnage and on the body colouration which would make it a particularly attractive fish to aquarists. The body appears slightly compact; the dorsal part is grey brown. The dorsal fin is low and the membranes do not exceed the fin rays. In their posterior part, the dorsal and anal fin are not much extended. Like the females, the males also have only a relatively small caudal fin. The relatively short ventral fins show an orange tone, their anterior edge is whitish blue. Behind the eyes, a narrow dark horizontal band begins which ends already in front of the vertical oval dark caudal blotch with a slightly grey red blotch. Above this horizontal band, there are three bright blue scale rows whose identical clearances show lilac tones.

The spotted pattern of the gill covers adapts to the bright blue and lilac tones mentioned before. The rounded caudal fin shows six semicircular rows of lilac dots. On the other hand, in the clearances bright blue tones shimmer. The breast is yellowish, the posterior lower part of the body is tender grey.

<u>Total length:</u> Male 5 cm, female 4 cm.

Biotope: Rio Parnaíba, Piauí (Brazil)
upstream the city of Teresina
Date: 26.11.1996
Time: 10.30 o'clock
Water movement: In the central part of the river strong, near the banks still flowing; side pools were stagnant
Water colour: In the river cloudy and brown; in the pools clearer
Surface temperature: 28.4°C (banks) to 29.3°C (pools)

<u>**Laboratory results:**</u>

pH-value	6.58
Electrical conductivity (25°C), µS/cm	32
Acid capacity up tp pH 8.2 (p-value), mmol/l	n.p.
Acid capacity up to pH 4.3 (m-value), mmol/l	0.27
As carbonate hardness °d	0.76
Total Alkalis, mmol/l	0.106
As total hardness °d	0.59
Sodium (Na^+), mg/l	1.5
Potassium (K^+), mg/l	1.2
Calcium (Ca^{2+}), mg/l	2.61
Magnesium (Mg^{2+}), mg/l	1.0
Total Iron (Fe^{2+}/Fe^{3+}), mg/l	0.41
Manganese (Mn^{2+}), mg/l	0.008
Ammonium (NH_4^+), mg/l	0.05
Nitrite (NO_2^-), mg/l	<0.02
Nitrate (NO_3^-), mg/l	0.3
Chloride (Cl^-), mg/l	1.4
Sulphate (SO_4^{2-}), mg/l	0.9
Phosphate (PO_4^{3-}), mg/l	0.06
Hydrogen Carbonate (HCO_3^-), mg/l	16.5
Oxidation	
a) As $KMnO_4$ consume, mg/l	5.2
b) As Oxygen (O_2), mg/l	1.3

n.p. = not provable

S03800-4 *Apistogramma piauiensis*, threatening adult male above bright ground.

Maintenance in the aquarium and reproduction: With this example, it can be seen well that they adapt their own colouration to the bottom they are kept upon. Upon bright sandy ground, their colouration generally gets brighter. In courtship colouration, the dark horizontal band and the caudal blotch vanish completely: Then the male shines only in the pastell colours orange, bright blue and red. At a temperature of 26°C and water readings around 6.5 pH and 170 µS, the specimens spawned willingly. A peace of savannah wood with two cave-like interconnected hollows was selected as spawning place. In the left cave, about 60 eggs were attached to the ceiling with their poles. The entry was closed with sand up to a narrow crack. During the following 9 days,

S03800-4 *Apistogramma piauiensis*, female with young.

the female ›excavated‹ untiringly the fine sand out of the second cave which was used only as entry. Ten days after, the young were swimming free and in the following six nights they were regularly brought back into the protective cave by the female.

While keeping *A. piauiensis* upon fine quartz sand, it could be clearly stated that the specimens feel comfortable with this material and it seemed as if they would enjoy chewing the fine sand and ejecting it through the gills. These obser-vations are best known to most Cichlid keepers, whereas here it shall be reminded of the associ-ated group of *Geophaginii*. This should be con-sidered for the choice of the ground material.

Apistogramma pleurotaenia (REGAN, 1909)

Meaning of the specific name (etymology): From the Greek ›*pleuro*‹ = (body) ›side‹ and the Latin ›*taenia*‹ = band: ›with a lateral band‹. This refers to the lateral band which indeed nowadays is known in most of the representatives of the genus.

First description: Ann. Mag. Nat. Hist. (8) 3.

Synonyms: *Heterogramma pleurotaenia*.

Remark: In the beginning of this century, the merchant and aquarist from Hamburg, Paul ARNOLD, had many contacts to sailors who helped in potential imports of ornamental fish. The type material of this species (unfortunately only two females) which was first (officially) imported by Kittler (1905), came from him.

Distribution: In the book ›Fremdländische Süßwasserfische‹ (1936), the authors ARNOLD & AHL mention a distribution range all over »the entire South America, from the Amazon River in the north to the La Plata drainage in the south«. It is not certain if this knowledge is verified by findings. Such a wide distribution area has to be doubted. Besides the type material which was available to REGAN for his description and which is very limited in numbers, later further specimens were introduced. The ›chequered Dwarf Cichlids‹ shown on a photo in this book are specimens in a photo aquarium which clearly remind us of *A. borellii*, a species which at that time was also already scientifically described (REGAN, 1906), but information in the book is missing. Further data: »... reaches a length of 7 cm and is quite variable in pattern and colouration. Depending on southern or northern origin, the form also varies«.

Habitat: If one trusts in the information given by REGAN (1909) and those in SCHMETTKAMP (1982) and RICHTER (1987) which probably refer to the author of the first description, these representatives of *A. pleurotaenia* would be the southernmost occuring of all Dwarf Cichlids. In HOLLY, MEINKEN & RACHOW (loose leaf edition), a work in several volumes, in which MEINKEN worked upon the genus *Apistogramma*, the authors writes about the distribution of *A. pleurotaenia*: »Not rare in the entire drainage of the La Plata, from southern Brazil through Uruguay, northern Argentina, Paraguay up to Bolivia in quiet parts of rivers rich in wood and plants, as well as with stones and cliffs, but not easy to capture; the same is true for the Amazon area.« First of all a correction: Although the Gulf of La Plata on some maps (also Argentinian ones) has the name ›Rio de la Plata‹, this is not at all a big river, but rather a gulf, a part of the sea where the Rio Uruguay as well as the Rio Paraná enter. Does the authors mean the latter? RICHTER shows specimens of the species (!) which due to their type could ›fit‹. Who knows?

Even if already in 1911, HASEMAN took the view that *A. pleurotaenia* should be considered a synonym of *A. borellii*, KULLANDER has no doubt that these are two different species.

SD and colouration: In recent years, various books about Dwarf Cichlids showed specimens under this name whose valid membership to this species has to be doubted (for example RICHTER, 1987, page 99). We show photos of possible specimens of this species under the (temporary) popular name *A.* spec. ›Paraguay II‹.

Apistogramma pulchra KULLANDER, 1980

Meaning of the specific name (etymology): ›Beautiful‹.

First description: Bonner zool. Monogr. (14), pages 135-138.

Remark: The type material comes from a collection which was compiled by the former Belgian King Léopold III. together with the Belgian ichthyologist Dr J.-P. GOSSE in 1967.

S03800-4 *Apistogramma pulchra*, adult male.

Synonyms: None.

Distribution: The type locality is the small Rio Preto (= black river), a left sided tributary of the Rio Candeias. This one, coming from the south, passes Porto Velho on the east, the capital of the Brazilian federal state of Rondônia and north-east of this city it flows into the big Rio Madeira. Certain places at the Rio Candeias are for the inhabitants of Puerto Velho (cp. map page 41) popular for outings or holidays.

Habitat: These are the shallow parts near the banks of small flowing waters as well as rest waters and pools in their inundation zones. Here the Dwarf Cichlids live under the shelter of foliage and dead wood. MAYLAND travelled for the last time, in 1996, to the drainage of the Rio Candeias and thereby he also took the material for the water samples added herein.

Remark: In recent years, due to the interpretation of different traits, experts tended to place *A. pulchra* into the *agassizii* complex. But this did not correspond to the remarks in KULLANDER's description. In 1996, after another revision - also with life material - RÖMER (pers. comm.) stated the identity with *Apistogramma* spec. ›Porto Velho‹. On the base of these new discoveries, the classification into the *pertensis* complex can be seen also in an optical way and so corresponds to KULLANDER's findings.

SD and colouration: The body of *A. pulchra* is elongated. The dorsal fin is high, but it does not reach the heigth of the dorsal of *A. pertensis*. It can almost extend to the end of the caudal fin. The anal fin is small and pointed. The upper and the lower ends of the concave caudal fin have tips. Following RÖMER, in very old males

S03800-4 *Apistogramma pulchra*, yound sexually mature female in front of the breeding cave.

Biotope: Rio Candeias, near Porto Velho,
Rondônia (Brazil)
Date: 10.11.1996
Time: 10.30 o'clock
Water movement: Rapidly flowing
Water colour: Clear, greenish if seen from above
Surface temperature: 26.9°C

Laboratory results:

		Magnesium (Mg^{2+}), mg/l	0.23
		Total Iron (Fe^{2+}/Fe^{3+}), mg/l	0.17
pH-value	6.76	Manganese (Mn^{2+}), mg/l	0.003
Electrical conductivity (25°C), µS/cm	15	Ammonium (NH_4^+), mg/l	<0.02
Acid capacity up to pH 8.2		Nitrite (NO_2^-), mg/l	<0.02
(p-value), mmol/l	n.p.	Nitrate (NO_3^-), mg/l	0.4
Acid capacity up to pH 4.3		Chloride (Cl^-), mg/l	0.3
(m-value), mmol/l	0.12	Sulphate (SO_4^{2-}), mg/l	0.4
As carbonate hardness °d	0.34	Phosphate (PO_4^{3-}), mg/l	<0.04
Total Alkalis, mmol/l	0.030	Hydrogen Carbonate (HCO_3^-), mg/l	7.3
As total hardness, °d	0.17	Oxidation	
Sodium (Na^+), mg/l	0.66	a) As $KMnO_4$ consume, mg/l	6.9
Potassium (K^+), mg/l	1.5	b) As Oxygen (O_2), mg/l	1.7
Calcium (Ca^{2+}), mg/l	0.81	n.p. = not provable	

in the central part of the caudal fin some rays should show extended rays. The ventral fins are short.

In female specimens, the dorsal fin is less high, it is pointed but not extended. The caudal fin is truncated and without filaments. A thinner lateral band starts behind the eyes and ends clearly in front of the base of the caudal fin. An intense irregular rectangular blotch is partly on the base of the caudal peduncle as well as on the base of the caudal fin. An intense cheek stripe runs diagonally on the golden gill covers. Depending on the mood, in specimens of both sexes; a lateral blotch can be seen.

The anal and the ventral fins of the males are bright blue, whereas the dorsal and caudal fins are lilac. The caudal shows six vertical rows of golden spots. Concerning the conspicuous colouration, the elongated body as well as the finage, *A. pulchra* can be seen as a true representative of the *pertensis* complex.

Total length: Old males 7 cm, females clearly smaller.

Maintenance in the aquarium: Keeping *Apistograma pulchra* should be similar to as described in *A. pertensis*.

Reproduction: At present, the authors have no information yet about the reproduction of these specimens. But *A. pulchra* could be considered one of the more difficult species.

Apistogramma regani KULLANDER, 1980

Meaning of the specific name (etymology): Dedication name in appreciation and honour of Dr Charles Tate REGAN (1878-1943) and his extensive revision of the American Cichlids from 1905 to 1906. In 1913, he created the genus *Apistogramma*. He was the world famous British ichthyologist who worked at the British Museum of Natural History from 1901 to 1938, and left the Museum as Director of the Department of Zoology.

First description: Bonner zool. Monograph. (14), pages 66-72.

Synonyms: In former decades, specimens of the present *A. regani* were repeatedly erroneously kept under the name of species that were already scientifically described or - together with these - placed in collections: *Heterogramma ortmanni* (partly) HASEMAN, 1911 (not EIGENMANN, 1912); *Heterogramma taeniatum* (partly) DE MIRANDA RIBEIRO, 1918 (not GÜNTHER, 1862); *Apistogramma taeniatum* MARLIER, 1967 (not GÜNTHER, 1862); *A. borellii* MEINKEN, 1971 (not REGAN, 1906).

Remark: The type material comes mainly from a collection compiled by the former Belgian King Léopold III. together with the Belgian ichthyologist Dr Jean-Pierre GOSSE during their trips in 1967. Further legates come from collections of J.D. HASEMAN 81913), J. MCCORMICK (without date), G. MARLIER (1964) and R. GEISLER (1967).

Distribution: All the type material comes from the surroundings of the Anavilhas Island Group in the Rio Negro as well as a small Váreza lake (Váreza = inundated lowland) filled with white water, the Lago Redondo, which is on the right side of the Solimões 20 to 25 km southwest of Manáus. Remember: Solimões is the Brazilian (!) Amazon above the entry of the Rio Negro. RÖMER (1991[b]) found specimens of this species in a collection from an area northwest of the city of Santarém (cp. map page 59).

Detail of the Anavilhanas-island archipelago in the Rio Negro, the habitat of many Dwarf Cichlids.

Habitat: *A. regani* can be mainly found in the shallow areas where they occur in large numbers near the banks. Here they live near foliage covering the sandy or sedimentary bottom. The collector who steps into the bank areas in the Anavilhas Island Group that are covered with water, will see that the water only remains ›negro‹ (= black as coffee) as long as the sedimentary ground is not stirred up. Then the coffee-like tone is mixed up and turns into a beige that is like ›milk coffee‹. As the river is poor in nutrients and thus practically contains no insect larvae (the lack of mosquitos makes a stay in uninhabited areas a pleasure!), but as all the year round it contains freshwater crayfish in many species and ages, the Dwarf Cichlids

S03810-3 *Apistogramma regani*, young sexually mature male.

living here seem to feed mainly from this food. As can be seen in the water sample on page 100, the water is extremely soft, completely free of nitrate and clearly acidic.

SD and colouration: The body height can be considered average. A dark loamy basic colouration is on the head and the body which also reaches into parts of the dorsal fin. Beneath this, there is a fine bluish shimmer.

Also the slightly rounded caudal fin as well as the anal fin shimmer in a blue tone and additionally are covered with a pattern of rows of brown dots. In male specimens, the ventral fins are yellowish, in females the anterior fin rays are black and the following zone is whitish to yellowish. Dorsal blotches and vertical bars are occasionally shown. Then, the dark lateral band is dissolved in darker squares. Four rows of stripes on the lower part of the belly remain constantly visible.

The cheek stripe and the caudal blotch show the deepest black colour. On the cheeks there is a pattern of red and bright blue spots and lines. In the females, the basic colouration appears brighter and the pattern of the vertical bars is continuously more intense. The unpaired fins are transparent.

In courtship and breeding mood, the yellow basic colouration increases and the well known markings (anterior rays of the dorsal and ventral fins) show their intense black colouration.

In his work of 1980, KULLANDER refers to the differences in colouration in specimens from different localities.

Total length: Male 6 cm, female 4 cm.

Maintenance in the aquarium and reproduction: A member of the genus from central Amazonia which in maintenance causes relatively few problems if certain basic conditions in water quality are given: total hardness as soft as possible (4 to 6°d) and a slightly acidic pH value below the neutral point 7.0. The water temperature should be at 27°C as usual. The aquarium is furnished with caves made of wood and stones as well as with plants. For reproduction, the data given above has to be slightly ›improved‹ which means that the pH value has to be lowered slowly until it is clearly acidic, just as this can be found in the area of the Anavilhas Island Group (cp. water sample page 100). Reproduction itself then runs as often described in this book.

Apistogramma resticulosa KULLANDER, 1980

Meaning of the specific name (etymology): Adjectival form of the Latin ›*resticulus*‹ = ›with fine lines‹ - shall refer to the lines on the lower part of the belly (which in living specimens can not always be seen).

First description: Bull. Zool. Mus. Univ. Amsterdam 7 (16), pages 157-164.

Synonyms: None.

S03815-3 *Apistogramma resticulosa*, sexually mature male.

Distribution: Until today only known from the area of the type locality, where in August 1976 the type material was collected by the US American editor Herbert R. AXELROD and the French ichthyologist Dr Jacques GÉRY.

Until today, the distribution range is considered a small river (Rio Xicanga) 18 km west of the city of Humaitá at the Rio Madeira. The city is near the crossroad of the ›Transamazônica‹ Porto Velho - Manáus and the ›Rodovia Transamazônica‹ coming from Santarém (cp. map page 56).

Habitat: An area where there is mixed water which at a high level can be muddy due to the overflowing water of the Rio Madeira which is rich in sediments. Here the water readings are extremely soft (total and carbonate hardness <1°d), nitrate is absent (<1) and the pH value is about 6.5 and 6.0. In areas with rich rainforest, the sandy bottoms are covered with twigs, coarser branches as well as with foliage.

SD and colouration: The specimens are not particularly colourful, but if they are relaxed, their silvery grey to olive body is covered by a sky blue glimmer (photo). On the cheeks, one can see a pattern of dark red lines and dots.

The trivial name which was introduced before the scientific description was therefore ›Wangenfleck-Apistogramma‹ - but other Dwarf Cichlids also show red blotches on the cheeks, such as *A. caetei* and *A. geisleri*. The fins are relatively transparent and also shimmer bluish. In specimens of both sexes, the caudal fin is rounded. In the general appearence, females have a more intense yellow colouration.

These specimens mostly show a more intense black pattern on the body and the fins. The cheek stripe, in males often suppressed, can be seen in femals as clearly as the vertical oval caudal blotch. During courtship and breeding time, the yellow tones in the females increase in intensity and the anterior rays of the dorsal and ventral fins develop to deep black.

Total length: Male 6 cm, female 4 cm.

Maintenance in the aquarium and reproduction: The specimens are kept in an aquarium of between 60 and 80 cm. Sandy bottom respectively very fine gravel, caves made of roots and stones as well as an opulent background vegetation provide them an almost normal habitat. For reproduction, the water should be as soft as possible, clearly acidic and as poor in nitrate as possible. The temperature is adjusted to the common 27°C. Reproduction occurs, following the principle of male-mother-family, frequently described in this book. In about four days the eggs develop up to the hatching of the larvae. After repeated transfers by the mother, the young fish swim free after some more four to five days and they immediately consume *Artemia* nauplii.

In 1987, SUTTNER described some successful breeding attempts in water whose total hardness was about 150 µS and the pH value was only 6.9. The water temperature was adjusted to 28°C. The author describes the different behaviour of the same couple - particularly of the female - in two tanks with different size and equipment. »This new behaviour, created by the change of biotic (for example swimming plants, opponent fish) and abiotic factors (for example strong current, larger water volume) was very surprising to me.« It shows that in these fish often more interesting behaviour can be observed than is at first supposed.

Biotope: Rio Xicanga,
near the city of Humaitá, Amazon (Brazil)
Date: 9.11.1996
Time: 13.30 o'clock
Water movement: Near the banks still to moderately flowing
Water colour: After a rainfall slightly coloured.
Surface temperature: 28.9°C

Laboratory results:

pH value	5.92
Electrical conductivity (25°C), µS/cm	17
Acid capacity up to pH 8.2 (p-value), mmol/l	n.p.
Acid capacity up tp pH 4.3 (m-value), mmol/l	0.11
As carbonate hardness, °d	0.31
Total Alkalis, mmol/l	0.041
As total hardness, °d	0.23
Sodium (Na^+), mg/l	0.70
Potassium (K^+), mg/l	1.3
Calcium (Ca^{2+}), mg/l	0.72
Magnesium (Mg^{2+}), mg/l	0.56
Total Iron (Fe^{2+}/Fe^{3+}), mg/l	0.58
Ammonium (NH_4^+), mg/l	<0.02
Nitrite (NO_2^-), mg/l	<0.02
Nitrate (NO_3^-), mg/l	2.7
Chloride (Cl^-), mg/l	0.6
Sulphate (SO_4^{2-}), mg/l	0.3
Phosphate (PO_4^{3-}), mg/l	<0.04
Hydrogen Carbonate (HCO_3^-), mg/l	6.7
Oxidation	
a) as $KMnO_4$ consume, mg/l	15
b) as Oxygen (O_2), mg/l	3.8

n.p. = not provable

Apistogramma ritensis (HASEMAN, 1911)

Meaning of the specific name (etymology): This is the adjective form of ›Rita‹. The name refers to fact that the specimens were found in the Rio Santa Rita.

First description: Ann. Carn. Mus. 7, page 362.

Synonyms: *Heterogramma ritense.*

As distribution range The author mentioned only a 2.5 cm long specimen ("Cotypes are 2 to 2.4 cm long"). Type locality is the Rio Santa Rita »of the Paraguay basin«.

Remark: The author (HASEMAN) remarks: »The coloration, shorter and deeper body, fewer scales, gill-rakers and shorter caudal peduncle separate this species from its nearest ally, *H. borellii.*«

KULLANDER (1980, page 25) sees the species »or considered synonyms on basis of careful consideration of the original description« as synonym of *A. borellii.*

The numerous vertical bars (8 to 9) shown by the specimens on an added sketch (which was also copied by FOWLER [1954]) allowed one to see that this is a specimen from the affiliation of *A.* spec. ›Paraguay‹ which is mentioned herein.

Apistogramma rondoni

(A. DE MIRANDA RIBEIRO, 1918)

Meaning of the specific name (etymology): A dedication name in honour of General Cándido Mariano da Silva RONDON (1865-1958).

In 1939, he became the first president of the newly founded Brazilian Indian Reservation Council. In 1952, he presented to the then state President Getúlio Vargas a draft bill for the foundation of the National Park Xingú for the protection of flora and fauna of that area, and for the conservation of the habitat of the locally living natives. Due to his cause, in 1953 the ›Museu do Índio‹ of Manáus was founded. As a pioneer and protector of the Indians, he received numerous honours in Brazil and other countries. For example, in 1953 a city in the Mato Grosso received the name ›Rondonópolis‹. In 1955, the Brazilian Congress raised him to the rank of a Marshal. In 1957, the former federal territory Guaporé which Rondon examined during many of his trips, was named ›Rondônia‹ (today a federal state) in his honour. In the same year, he was nominated for the Nobel Peace Prize.

First description: Comm. Linhas elegr. Matto Grosso ao Amazonas Publ. 46 (5), pages 16-17.

Synonyms: *Heterogramma rondoni.*

Distribution: Described from the Rio Paraguai (Brazilian name for the Rio Paraguay, as it is called in Paraguay and Argentina) in the surroundings of the city of Cáceres in the Brazilian federal state of Mato Grosso. These could (?) be specimens which are herein mentioned and shown in photos under the popular name *A.* spec. ›Paraguay I-III‹.

Remark: The type material which had been deposited for the description could not be examined (lost?) so that a more exact identification with the specimens of the type ›Paraguay‹ mentioned here could hardly be possible.

Apistogramma roraimae KULLANDER, 1980

Meaning of the specific name (etymology): From the northern Brazilian federal state of Roraima, bordering to Venezuela.

First description: Bonner zool. Monograph. (14), pages 138-141.

Synonyms: None.

As distribution range the greater surroundings of Boa Vista, the capital of Roraima, at the upper Rio Branco is given.

Habitat: Water readings as for *A. rupununi.*

Remark: As we know today, and as the author (KULLANDER) verified, this description is an error as the examined material turned out to belong to *A. gibbiceps.*

Apistogramma rupununi FOWLER, 1914

Meaning of the specific name (etymology): Named after the Rupununi River in the southwest of Guyana, the former British Guyana. The Rupununi is a tributary of the Essequibo River.

First description: Proc. Ac. Nat. Sci. Philadelphia 66, pages 277-278.

Synonyms: *A. ortmanni rupununi* EIGENMANN, 1912.

Remark: In his revision from 1980, KULLANDER considered *A. ortmanni rupununi* FOWLER, 1914 as a synonym of *A. steindachneri.* In 1989 (page

S03820-3 *Apistogramma rupuruni*, sexually mature male.

82), the author explains that a new revision of the type material (two adult males as holo- and paratype as well as a larger number of preserved specimens from the surroundings of Boa Vista [Roraima/Brazil] makes obvious that indeed *A. rupununi* is an independent and valid species. The species which was formerly named *Apistogramma* spec. ›Zweifleck‹ in German magazines and books, differs from the related *A. hippolytae* particularly by a more narrow lateral blotch which only reaches into the upper lateral line instead of the basis of the dorsal fin.

Distribution: From the drainage of the Rupununi River in the south-west of Guyana through the watershed towards Brazil, to the south up into the drainage of the upper Rio Branco (Boa Vista) in the Brazilian federal state of Roraima. In 1989 (page 82), KULLANDER refers to the fact that the data in KANAZAWA's description (1966) of *A. rupununi*, the river with the same name and the correlated coordinates (2-3°N, 50°20'W) had to be a mistake because (50°20'W) would be the latitude of the Ilha de Maracá upstream the delta of the Amazon River. This is probably a reading error, whereby the ›50‹ should be replaced with a ›59‹ (cp. map page 141).

Habitat: Depending on the habitat, the course of the stream or the inlet, the water can show different turbidity. Water readings like those of the upper Rio Branco can be seen from the

Biotope: Rio Branco, (upper part) at the banks near the city Boa Vista/Roraima
Date: 6.10.1986
Time: 7.00 o'clock
Water movement: fast flowing
Water depth: Not established
Water colour: Moderately cloudy from sediment

Laboratory results:

pH-value	6.30
Electrical conductivity (25°C), µS/cm	23
Acid capacity up to pH 8.2 (p-value), mmol/l	<0.05
Acid capacity up to pH 4.3 (m-value), mmol/l	0.26
As carbonate hardness, °d	0.7
Total Alkalis (Ca^{2+} + Mg^{2+}), mmol/l	0.07
As total hardness, °d	0.4
Calcium (Ca^{2+}), mg/l	1.5
Magnesium (Mg^{2+}), mg/l	0.70
Total Iron ($Fe^{2+/3+}$), mg/l	0.12
Manganese (Mn^{2+}), mg/l	<0.02
Ammonium (NH_4^+), mg/l	0.06
Nitrite (NO_2^-), mg/l	<0.02
Nitrate (NO_3^-), mg/l	<1
Chloride (Cl^-), mg/l	<3
Sulphate (SO_4^{2-}), mg/l	8
Total Phosphate (PO_4^{3-}), mg/l	0.06
Hydrogen Carbonate (HCO_3^-), mg/l	16
Oxidation Potassium Permanganate $KMnO_4$ consume, mg/l	321

listing below. As it is to assume that the rainforest in the south-west of Guyana is still intact, also here one can suppose the water conditions that are common in the Amazon - even though the rivers have an almost opposite direction. Here, wood and foliage should also cover the sandy or rocky bottom of the shallow parts near the banks of the rivers.

The Gran Sabana, whose extentions stretch from Venezuela into Guyana, send their very soft and absolutely nitrate free water (total and carbonate hardness <1°d, electric conductivity between 5 and 15 µS) partly towards the drainage of the Essequibo (cp. also water sample of the Rio Apónguao in *A. ortmanni*).

SD and colouration: One of the first photos of the true *A. rupununi* was presented by SCHMETTKAMP (1982), whereas in KOSLOWSKI (1981) as well as in LINKE & STAECK (1984/1. Edition) it was a second photo of *A. hippolytae*.

Head and body as well as dorsal, caudal and ventral fins are brass. The ventral part remains a little bit brighter. Normally, the cheek and lateral band - even depending on the mood - are shown. This is also true for the caudal blotch at the end of the caudal peduncle. The blotch on the central upper half of the flanks that has already been mentioned ›sits‹ upon the lateral band and ends a few millimeters below the top of the back, i.e. at the upper lateral line. The caudal fin is rounded in both sexes and covered by a pattern of vertical rows of dark dots. They stretch into the soft parts of the dorsal and anal fin. The anterior part of the anal fin shimmers bluish.

Total length: Males can reach up to 6 cm, but females only 4.5 cm.

Maintenance in the aquarium and reproduction: Unfortunately, little is known about the maintenance in the aquarium, and above all also about reproduction.

Apistogramma spec. ›Alto Rio Negro‹ or ›São Gabriel‹

New scientific name:

Author and year:

Remark: Until today, the specimens were exclusively presented by MAYLAND (1987). For the first time, they were captured by BLEHER in October 1986 some eight to ten kilometers north of the town of São Gabriel da Cachoeira at the upper Rio Negro; in a small gully with black water behind the outlet of a culvert made of corrugated iron.

Distribution: Only known from this area (cp. map page 121).

Habitat: See above: water 26°C, electric conductivity 20 µS, pH value 5.2).

SD and colouration: See the two photos (both specimens in breeding colouration). A bright violet zone on the caudal peduncle of the males is conspicuous.

Total length: Male 5 cm, female 4.5 cm long.

Maintenance in the aquarium and reproduction: Until today, only a few mentioned specimens were caught and imported to Germany for reproduction. The only offspring was offered in the trade by BLEHER's company (Aquarium Rio) without success. Later the semi-adult offspring was exported to Japan.

S03838-4 *Apistogramma* spec. ›São Gabriel‹, male and female, the male specimen displays the blue violet colouration during the breeding period only.

S03846-4 *Apistogramma* spec. ›Blutkehle‹, head portrait.

Apistogramma spec. ›Blutkehle‹
New scientific name:
Author and year:

Distribution: The specimens coming from the greater surroundings of the eastern Colombian border town of Puerto Inirida at the Orinoco and are mostly imported occasionally via Bogotá together with other fish. Often they are found with specimens of *A. iniridae* which are equally imported occasionally, obviously in the habitats they occur together.

Occasionally, a few specimens of both species are imported together with *Dicrossus filamentosus* (cp. map page 121).

Habitat: Similar to as described in *A.* spec. ›Breitbinden‹.

S03846-4 *Apistogramma* spec. ›Blutkehle‹, adult male.

S03846-4 *Apistogramma* spec. ›Blutkehle‹, female in breeding colouration.

SD and colouration: This species was for the first time presented by SCHMETTKAMP (1981) and owes its trivial name to the pattern of blood red stripes which - particularly in an aggressive mood - can be seen on the throat. Additionally, then the mentioned part becomes intensely red. The specimens are not as slender as *A. iniridae*, and the males can be differentiated from these particularly well by their very big bilobated caudal fin. Stripes on the snout as well as the cheek and lateral band are well pronounced and are shown depending on the mood. The males can reach an impressive total length and their fins do not take second place in their relative size. Also this dorsal fin (just like in *A. iniridae*) is sail-like, whereby the membranes exceeding the spines are grown together all over the length of the fin.

Concerning the females, SCHMETTKAMP remarks: »Young females of *A. iniridae* and the ›Blutkehl-Apistogramma‹ can hardly be differed from each other. Adult females of the ›Blutkehl-Apistogramma‹ develop a crust like bilobated caudal fin and then they can be separated from the round-tailed females of *A. iniridae*. A differentiation of females of the ›Rotkeil-Apistogramma‹ is even more difficult because these females either have the bilobated form of the caudal fin. But in breeding mood, the females of the ›Rotkeil-Apistogramma‹ become red orange and only now they be differentiated from the females of the ›Blutkehl-Apistogramma‹ which stay yellow«.

Total length: Male 10 cm, female 7 cm.

Maintenance in the Aquarium and reproduction: As the specimens come from water that is very poor in minerals and additionally is clearly

S03848-4 *Apistogramma* spec. ›Breitbinden‹, rivaling males.

S03848-4 *Apistogramma* spec. ›Breitbinden‹, adult male.

acidic and warm, one can recommend a breeding water with a total hardness of 3°d, a pH value around 5.5 and a water temperature of 28°C. Corresponding to their size, the males occupy a relatively wide territory in an aquarium that should be longer than 80 cm.

In this concern, SCHMETTKAMP, who was keeping the specimens and was trying to reproduce them, remarks: »Reproduction seems to be extremely difficult and requires extreme water readings. At a hardness around 1°dGH and a pH of 5.5, they spawned, but the eggs did not develop«. At water readings of 3°dGH and a pH value of 6.5, Bork could also repeatedly watch the specimens spawn (and fertilize?). But even after repeated spawning, the eggs did not develop.

Apistogramma spec. ›Breitbinden‹

New scientific name:

Author and year:

Distribution: The specimens are mostly imported by importers together with other fish. They mainly come from areas at the upper Rio Orinoco (border between Venezuela and Colombia) as well as at the Rio Atabapo and are imported from Bogotá together with *Pterophyllum altum*, *Paracheirodon axelrodi* and other ornamental fish species. RÖMER (1995) reports about catches

S03848-4 *Apistogramma* spec. ›Breitbinden‹, female with young.

★ *A. spec. „Alto Negro“*
△ *A. spec. „Breitbinden“*
▲ *A. spec. „Blutkehl“*

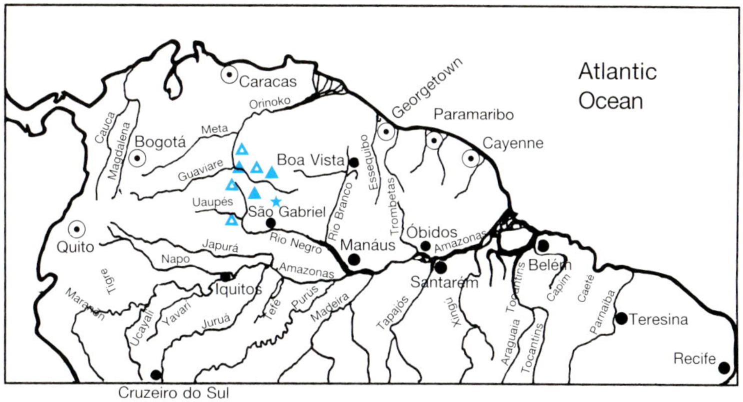

of this species in the Rio Uaupés, so that the distribution in the area of the upper Rio Negro would also be proved.

In 1996, some shipments were introduced whose content was erroneously declared as ›*A. guttata*‹. But these were up to 100% young sexually indistinguishable specimens of *A.* spec. ›Breitband‹. Hence, these cannot originate from the Rio Morichal (isolated occurrence of *A. guttata* north of the Orinoco delta), but they probably come from an area near the town of Puerto Ayacucho at the central Orinoco (cp. map page 121).

Habitat: Here the water is extremely soft (total and carbonate hardness <1 °d, electric conductivity 5 to 10 µS) and clearly acidic. All around, there is still plenty of rainforest so that one can assume that the sandy beds of the streams at least near the banks contain considerable wood and foliage.

SD and colouration: Adult males show a silvery basic colouration on the head and the body. A relatively wide horizontal band runs through the flanks and ends in a small black caudal blotch. These cheek stripes are shown only depending on the mood. On the cheeks, one can see a row of little blood red blotches. The highly looming dorsal fin whose membranes are all well extended, are reddish. Equal tones can also be found in the peripheral parts of the bilobated caudal fin whose centre additionally shows a pattern of red brown dots.

In courtship and breeding mood, the smaller females are yellow all over the head and body. Their dorsal fin does not loom, and also the caudal fin is only rounded. The cheek stripe and (mostly) two lateral blotches (one of which is bigger) shine deep black above the golden yellow of the head and body as well as the anterior rays of the dorsal and ventral fins.

Total length: Male 8 cm, female 6 cm.

Maintenance in the aquarium and reproduction: Even as the specimens come from an area with extemely soft water, for example reproduction was successful in water with a total hardness between 5 and 9°d. A lowering of the pH value below the 5.0-point was gained with the help of oak essence and constant control up to the hatching of the larvae. Additionally, ›Cilex‹ was added following the instruction, at first 100%, later only 50%.

In a tank of 80 cm length, the fish spawned inside a flower pot that was not too small and which had tiny entry. After the fertilization of about 120 eggs, the narrow entry was additionally closed with more sand. Occasionally, the female controlled the entry and chased away supposed enemies by short aggressive advances. The male itself was chasing troublemakers in the greater surrounding of the ›breeding cave‹. Thereby, also inhabitants of a neighbouring aquarium were attacked - as far as the separating screens permitted. At the common temperature of 27°C, the larvae hatched the fourth day, they were immediately transferred onto a nearby root, but soon after they were brought back into the flowerpot cave. This procedure was repeated several times. At that time, a ritualised scrubbing by both specimens at various substratum was noticed. Between eight and nine days after spawning, the young fish started swimming free. The mother was leading the offspring in an exemplary manner. The fry immediately consumed *Artemia* nauplii. Every day a minor water change with combination of previously prepared tap and osmosis water was done.

The young fish very much gravitated to the bottom and due to the good diet of *Artemia* nauplii, they grew and later they could develop astonishingly fast by additionally offered seived little *Daphnia*. Already after about four weeks, they had raised their total length to 15 millimeters.

Apistogramma spec. ›Gelbwangen‹
(without photo)

New scientific name:

Author and year:

Distribution: This species was first presented by Koslowski (1984). Often, the specimens are occasionally introduced via Manáus. More detailed information about their distribution range is not yet available (cp. map page 152).

Habitat: As the specimens at least live in central Amazonia, one can assume that they live in very soft and clearly acidic water. The nitrate concentration in these areas can almost not be detected. Also in the habitats of these Dwarf Cichlids in the still existing rainforest, wood and foliage is normally present.

SD and colouration: Adult males have a yellow grey basic colouration. More intense yellow tones can be seen in the peripheral parts of the head and body as well as in the fins, particularly on the cheeks, which helped the Dwarf Cichlids to acquire their German name. If light falls in with a diagonal angle, neck and body shimmer turquoise. Depending on the mood, the black pattern is visible in the form of a cheek and lateral band. The deep black caudal blotch is vertically oval. The posterior caudal fin is truncated.

In this species, not only the females show intensely black anterior dorsal and ventral fin rays, but also their male partners. In the females, whose yellow glimmer is clearly more intense during courtship and breeding time, then the horizontal band transforms into a stretched blotch pattern.

Total length: Male 6 cm, female 4.5 cm.

S03775-4 Unfortunately there are very few imports of these wonderful specimens of *Apistogramma* spec. ›Sunset‹.

S03906-4 *Apistogramma* spec. ›Orangesaum‹, wild male.

Maintenance in the aquarium and reproduction: Assuming that the specimens come from the black water habitats in the drainage of the upper Rio Negro, it is recommended to maintain and breed them as described for *A. regani*. In tanks, the speciemens were repeatedly reproduced and offered as well as sold in the trade or at fish sales under different popular names.

Apistogramma spec. ›Sunset‹

New scientific name:

Author and year:

Distribution: A new species scientifically undescribed. The first assumptions, that it could be *A. payaminonis* (DATZ, 1996), were not verified. The Peruvian (or Brazilian?) exporter named the locality Rio Yavari.

Remark: Due the physical appearance, this species has to be associated with the *cacatuoides* complex. *A. norberti* seems to be the closest relative. At the best, the rather inexact information about the locality admits a certain relationship to the species mentioned.

Habitat: As for the time being, more detailed information is missing, one can only assume a very similar habitat as it is known for *A. norberti* or *A. cacatuoides* (see there).

SD and colouration: As shown by the photo above, adult specimens (males) do not have a hefty neck part, but a rather flat upper head part. The wide black lateral band ends in almost square caudal blotch which stretches into the base of the caudal fin. Below this band, there is stripe-like sky blue zone running from the posterior edge of the gill cover to the caudal blotch. Ventral part and throat are intense golden yellow. The black cheek stripe is covered by green iridescent tones. A pattern of bright blue dots and lines is behind the equally golden eye ring.

The anterior five rays differ from the rest of the dorsal fin: The first two are decidedly shorter, whereas the rays numbering three to five, are extended and additionally turquoise iridescent. The turquoise colouration continues less intensely in the rest of the dorsal fin as a submarginal band. About sexual differences and the colouration of the females (particularly about their breeding colouration) nothing can be said yet because until today not enough specimens were imported.

S03906-4 *Apistogramma* spec. ›Orangesaum‹, wild female.

Total length: Corresponds to the specimens of *A. norberti* (see there).

Maintenance in the aquarium and reproduction: Information about maintenance and breeding are still missing because of a lack of imported specimens, but their care as well as breeding should be similar, as this is mentioned for the (supposed) relatives

Apistogramma spec. ›Orangesaum‹

New scientific name:

Author and year:

Distribution: The natural habitats of these specimens which are often imported occasionally with other fish, could not yet be investigated exactly. They probably come from the area of the upper Rio Orinoco in Venezuela or Colombia and were introduced together with Neon Tetras from Bogotá/Colombia (cp. map p. 181).

Remark: On pictures in aquaristic reports (cp. WINDISCH, 1991 ›Weißsaum-Apistogramma‹), specimens with similar seams can be seen, but these are not identical with the species presented here.

Habitat: The water in the area mentioned above distinguishes itself by a very low total hardness which with a common device can often not be detected. The pH value is clearly acidic.

SD and colouration: The basic colouration can be called golden brown. Depending on the mood, a horizontal band with lateral blotch situated in the centre of the body can be seen. Also depending on the mood, a black cheek stripe is shown. In specimens of both sexes the base of the caudal fin is limited by vertical brown band. The caudal fin is covered with semicircular rows of golden spots. The clearances are orange brown. Contrary to the bright blue anal fin with its orange brown pattern, the dorsal fin is slightly coloured orange; on the other hand, the tips of its membrane show a more intense colouration (name). The transparent ventral fins show a bright blue glimmer. Every gill cover is golden with a row of turquoise spots. In the males, the ventral part is a bit brighter than in the female partners. The latter can be distinguished from the males by the less pointed extensions of the anal fin, but particularly also by the clearly plump ventral area. Dorsal and anal fin are less opulently formed, which means that in the males indeed they are pointed, but they are not remarkably extended. The ventral fins are also only relatively short. The caudal fin is small and rounded. Probably this new species has to be placed into the *pertensis* group.

Total length: Male 3.5 cm, female a bit smaller.

Maintenance in the aquarium and reproduction: The conditions for maintenance and breeding could correspond to those of *A.* spec. ›Weißsaum‹. For the successful reproduction of this species, WINDISCH mentions a pH value of 4.5 as a condition.

S03907-4 *Apistogramma* spec. ›Pandurini‹, adult yellow female with cheek stripe.

Apistogramma spec. ›Pandurini‹

New scientific name:

Author and year:

Remark: Of the imports at first (mid of May 1996) consisting of single specimens (occasionally with other fish) which were considered a form *A. nijsseni*, a few months later complete small shipments could be organized. Meanwhile experts also confess that the ›Pandurini-Apistogramma‹ (named after the Peruvian exporter Jesus Victoriano Panduro Pinedo) is related to *A. nijsseni*, but until now a scientifically undescribed species (cp. map page 93).

Distribution: As far as known today, in an area on the right side of the Rio Ucayali which could follow or overlap with habitats of *A. nijsseni*.

S03907-4 *Apistogramma* spec. ›Pandurini‹, adult male during lateral threatening.

S03907-4 *Apistogramma* spec. ›Pandurini‹, brown female with cheek stripe.

S03907-4 *Apistogramma* spec. ›Pandurini‹ brown female with cheek blotch.

Habitat: Compare *A. nijsseni*.

SD and colouration: In full-grown males, the membranes of the first three dorsal fin rays are clearly extended, a fact that until now could not be found in types of *A. njisseni*, but - similar to *A. nijsseni* - in the colouration as well as in the pattern the ›Pandurini-females‹ are very variable. As a rule, on a splendid golden yellow body colouration the females show a deep black cheek stripe as well as an equally black, moderately wide vertical bar. The body part behind the vertical bar is covered with a slight green glimmer. The rounded transparent caudal fin has red edges. In particular mood - even if more rarely - on the caudal peduncle there is a blotch of moderate size which like a wedge, stretches into the usually transparent caudal fin. In males, this typical pattern can be constantly seen as a dark blue wedge. By this blotch, the females of the ›Pandurini-Apistogramma‹ preferably differ from those of *A. nijsseni*. In the latter ones, the dark caudal blotch is always round. Instead of the cheek stripe, some females show a big deep black cheek blotch. In these females, the equally black vertical bar widens itself in the ventral part - more or less wedge shaped - to a wide black zone (photo).

A third colour variety can be seen in those females in which the splendid golden yellow body colouration is replaced by grey brown one. Here the cheek stripe as well as the vertical bar are not so intense. In this variety, the wedge-shaped caudal blotch can also be seen only temporarily. The mentioned differences in colouration and pattern of the females show a clear parallel to *A. nijsseni*. Also in these specimens, the females

S03907-4 *Apistogramma* spec. ›Pandurini‹, yellow female displaying breeding colouration with young - 10 days old

are variable. Indeed, as mentioned the wedge-shaped caudal blotch in the females of *A. nijsseni* can never be observed. In fact, also in this species there are specimens of both sexes which show the mentioned grey brown colouration.

Total length: Male 7.5 cm, female 6 cm.

Maintenance in the aquarium and reproduction: Concerning maintenance and breeding, *A. nijsseni* and the ›Pandurini-Apistogramma‹ do not differ from each other. Fortunately, the reproduction in the aquarium of the latter has been achieved successfully many times. With water readings of about 150 µS and a pH value of 6.5 and a water temperature of 26°C, the specimens spawn mostly without problems. The females willing to spawn show their readiness with stormy courtship behaviour. The eggs are deposited with their poles at the ceiling of the cave. The eggs are cherry red. After about four days the equally cherry red larvae hatch. The care for the brood is very intense. If corresponding cover is present, in the coming days the brood is repeatedly transferred. Nine days after the fertilization, the young fish are for the first time lead around by the female in a dense swarm. Now also the young fish still show a clear red colouration. Newly hatched *Artemia* nauplii belong to the first food for the young fish and are immediately consumed. The further breeding happens as usual with the other *Apistogramma* species.

About the species from the small waters in the Rio Paraguay drainage:

The headwaters of the Rio Paraguay (in Brazil: ›Rio Paraguai‹) are in the Brazilian federal state Mato Grosso, upstream the city of Barra do Bugres (north-west of Cuiabá), where the river already recieves water from many smaller tributaries. On its stony course through the Pantanal, the southern federal state Mato Grosso do Sul and the bordering Paraguay, it forms a part of the border, then it cuts the neighbouring state Paraguay, passes its capital Asunción, from here it forms again a part of the border to the northern Argentina, in order to finally - not far away from the northern Argentinian cities of Corrientes and Resistencia - unify with the Rio Paraná coming from Brazil.

Although the Rio Paraguay was examined very intensely, recently from its drainage some interesting Dwarf Cichlids came to us which at first can not be well classified in the circle of the de-scribed species (if declared synonym or not). The small waters of the affluents of the Rio Paraguay in the mentioned countries possibly still contain further surprises.

Apistogramma spec. ›Paraguay I‹

New scientific name:

Author and year:

Distribution: Paraguay, in the drainage of the river with the same name (cp. map page 152).

Habitat: Relatively (not extremely) soft water (conductivity about 100 to 160 µS) with a constant neutral pH value (about 7.0). Many habitats have dense submersed vegetation and are additionally covered with swimming plants. Wood and/or foliage exists only where the environment is not yet cleared which means is still covered with forest.

SD and colouration: The dorsal fin of the relatively compact and high-bodied males is not raised much. Its end is extended with a filament

Biotope: Rio Paraguai, Mato Grosso (Brazil),
upstream the city of Cáceres
Date: 7.11.1996
Time: 13.10 o'clock
Water movement: Moderately flowing
Water colour: Clear, tea coloured, sandy ground
Surface temperature: 26.4°C

Laboratory results:

pH value	6.84
Electrical conductivity (25°C), µS/cm	39
Acid capacity up to 8.2 (p-value), mmol/l	n.p.
Acid capacity up to 4.4 (m-value), mmol/l	0.31
As carbonate hardness, °d	0.87
Total Alkalis, mmol/l	0.14
As total hardness, °d	0.78
Sodium (Na^+), mg/l	1.5
Potassium (K^+), mg/l	2.7
Calcium (Ca^{2+}), mg/l	3.1
Magnesium (Mg^{2+}), mg/l	1.5
Total Iron (Fe^{2+}/Fe^{3+}), mg/l	0.60
Manganese (Mn^{2+}), mg/l	0.004
Ammonium (NH_4^-), mg/l	<0.02
Nitrite (NO_2^-), mg/l	<0.02
Nitrate (NO_3^-), mg/l	3.1
Chloride (Cl^-), mg/l	1.1
Sulphate (SO_4^{2-}), mg/l	1.0
Phosphate (PO_4^{3-}), mg/l	0.08
Hydrogen Carbonate (HCO_3^-), mg/l	19
Oxidation	
a) as $KMnO_4$ consume, mg/l	13
b) as Oxygen (O_2), mg/l	3.3

n.p. = not provable

S03909-4 *Apistogramma* spec. ›Paraguay I‹, adult wild male.

and can reach the posterior edge of the rounded caudal fin. The equally extended and pointed anal fin has a similar length.

On the other hand, the ventral fins are short and wide. They show bright yellow tips. The upper part of the head and the back are coloured olive grey. The ventral part remains whitish and the lower part of the posterior body shows a bright grey colouration that reaches up to the base of the caudal fin.

The narrow cheek stripe gets narrower towards the lower edge of the gill cover. A deformed horizontal band ends shortly in front of the vertical oval dark caudal blotch. Behind and below the eyes, the head is covered with bright blue lines and dots.

S03909-4 *Apistogramma* spec. ›Paraguay I‹, wild female.

The membranes of the dorsal and anal fin show a lilac and bright blue pattern and also the rounded caudal fin show similar tones: It shows seven rounded rows of lilac spots. The clearances shimmer bright blue.

Female specimens are also very deep-bodied. The ends of their dorsal and anal fins are less pointed, which is also true for the strong but short ventral fins.

In colouration and pattern, the females are very similar to their male partners. In this concern, the rounded caudal fin is an exception. It has a clearly weaker pattern and colouration.

Occasionally, above the centre of the flanks a smaller almost point-shaped lateral blotch can be seen.

In this species, the sex can be seen quite late - after the beginning of sexual maturity.

Total length: Male 5 cm, female 4 cm.

Maintenance in the aquarium and reproduction: Similar to as given for *A. borellii*.

S03908-4 *Apistogramma* spec. ›Paraguay II‹, adult wild male.

Apistogramma spec. ›Paraguay II‹
New scientific name:
Author and year:
Remark: Of the specimens of this species, until today only one male came to Germany as an occasional import together with suckermouth catfish (*Otocinclus*) from Paraguay. The specimens are quite compact (cp. map pape 144).

SD and colouration: The basic colouration of the body is silvery white, whereby the dorsal part is darkened brownish-olive. From the ventral fins to the base of the caudal fin, the ventral part of the body appears to be covered with a pink shimmer. In its base, the otherwise clear rounded caudal fin is also iridescent and slightly pink. A similar tone can be seen in the

S03909-4 *Apistogramma* spec. ›Paraguay III‹, adult wild female.

S03909-4 *Apistogramma* spec. ›Paraguay III‹, adult wild male.

eye ring. The gill covers as well as the breast shine golden yellow. Below the eyes, the former ones are covered with bright blue lines and spots. From the upper lip to the eyes runs a short (!) dark brown line pattern which continues behind the eyes in double length (about 5 mm). A silvery line pattern of about 5 mm in length runs to the lateral line in an upward angle of about 40°. The spines of the dorsal and anal fin are black brown; the membranes shimmer silvery blue. Depending on the mood, starting in the back - more or less concisely - six dark brown vertical bars occur. In the ventral part, the colouration of these bars turns into a brownish pink in order to dissolve them.

Total length: Male 4 cm, female a bit smaller.

Maintenance in the aquarium and reproduction: Until now, a sufficient number of breeding specimens was not available.

Apistogramma spec. ›Paraguay III‹

New scientific name:

Author and year:

Remark: The import of the specimens shown here occurred in autumn 1995 occasionally together with *Apistogramma borellii* from Paraguay. Altogether it were six specimens.

SD and colouration: The basic colouration of the males is a bright pastell orange. All the fins are transparent and show a slight approach of the same pastell colouration. The dorsal fin has a tender bright blue seam and the anal fin shows a pattern of equally bright blue lines. Below the eyes, a bright blue spot pattern can be seen. It continues in three rows behind the eyes almost up to the base of the caudal fin. The ventral fins are short and wide (cp. map page 46).

Also outside the breeding time, female specimens are lemon yellow and their flanks are covered with a slightly emerald green glimmer. In specimens of both sexes an unclear orange caudal blotch is present.

Conspicuous in this species: neither in the dorsal fin, nor at the anterior edge of the ventral fins, nor near the lateral line, do females display any black colouration and the observer misses a cheek stripe or a cheek blotch. Also male specimens lack any black colouration in form of the horizontal band or a lateral blotch. A cheek stripe is absent.

Total length: Male 3.5 cm, female about 3 cm.

S03975-3 *Apistogramma* spec. ›Peixoto-Azevedo‹, adult wild male.

Apistogramma spec. ›Peixoto-Azevedo‹
New scientific name:
Author and year:
Distribution: This relatively unknown species was imported by Uwe WERNER from central Brazil and brought to reproduction. He reported about the expedition and catch in 1989. Unfortunately, the specimens meanwhile vanished from the market. By this the same happened to them that happened to species ›Alto Rio Negro‹ or ›São Gabriel‹ imported and reproduced by BLEHER and MAYLAND. Where is the attention of the Dwarf Cichlid keepers?

S03975-3 *Apistogramma* spec. ›Peixoto-Azevedo‹, adult wild female.

If one drives on the federal road no. 163 from the capital of the state of Mato Grosso to the north, about 90 km from the border to the following federal state of Pará one reaches a quite dirty and therefore rather mosquito infested city ›reigned‹ by golddiggers. In former times it was named Matupá, on newer maps it has the name ›Peixoto de Azevedo‹. To the north and north-east of the city, the river of the same name (Rio Peixoto-Azevedo) flows to the northwest. Its water is troubled by the intense gold washing activities and is continuously poisoned by quicksilver. For information: The river flows into the Rio Teles Pires which is also named Rio São Manuel. This river flows into the big Rio Juruena which can be considered a tributary of the Rio Tapájos (cp. map page 106).

The fish were captured near the town of Matupá in the quiet opulently planted shallow water of the Rio Peixoto-Azevedo, whose ground here was covered with foliage and rotting twigs. The water temperature was 24 to 25°C. The pH was a bit above 6.0 and the total hardness below 1°d. Werner reproduced the small Cichlids with a length of about 6 cm in harder water with a pH value of up to 7.8!

Apistogramma spec. ›Rio-Caura‹

(without photo)

New scientific name:

Author and year:

Distribution: The specimens were at first presented by STAECK (1991) and were personally collected by him in the drainage of the Rio Caura in Venezuela (Rio Tiquire, Rio Sipao).

Habitat: Besides the above mentioned distribution range, very similar fish were also found - together with *Microgeophagus ramirezi* and *Apistogramma hoignei* - in ponds surrounded by palms and partly used as watering places for cattle north of the Orinoco delta. The bottom of ponds as well as flowing waters are either sandy or covered with sediments and filled with mud. Aquatic vegetation can be present, but the fish can obviously get along without underwater vegetation. In such cases, they search for shelter under the overhanging vegetation of the banks or under foliage and wood accumulations. As almost all over Venezuela, the water is very soft, absolutely free from nitrate and more or less acidic.

SD and colouration: The colouration of the specimens seems to be limited. They are relatively deep bodied and - depending on the origin - the basic colouration of the males can reach from a sandy yellow to a silvery tone overlapped by blue. The pointed ends of the soft rays of the dorsal and anal fin are well extended and in adult specimens they reach up to the end of the slightly rounded (or truncated) caudal fin.

Female specimens tend toward beige tones, in which the black colour markings that are mostly typical for the genus clearly step forward. During the courtship and breeding time, they are coloured intensely yellow and then the mentioned black markings - here these are particularly the anterior dorsal and ventral fin rays - take over a deep black colouration. The horizontal band mostly appears only as a spotted band.

Total length: Male 7.5 cm, female 5.5 cm.

Maintenance in the aquarium and remark: See *A. hoignei*. Following some Dwarf Cichlid experts, this species should be a colour variety of *A. hoignei*. Examinations are still missing.

Apistogramma spec. ›Rotkeil‹

A colour form of *A. uaupesi* (see there).

Apistogramma spec. ›Rotpunkt‹

New scientific name:

Author and year:

Distribution: The natural habitats of these specimens that are mostly only imported occasionally together with other fish, could not yet be investigated exactly. They probably come from an area of the upper Rio Orinoco in Venezuela or Colombia. The affinity to *A. macmasteri*, which comes from the same area, would allow this hypothesis (an unproven suggestion) (cp. map page 134).

Remark: SCHMETTKAMP (1978) gave it its popular name. Recently, repeated imports followed by which specimens of *A.* spec. ›Rotpunkt‹ were mixed with those of *A. macmasteri*.

Habitat: Roughly corresponds to that of *A. macmasteri*. For years, this species is regularly reproduced by a few keepers. In reproduction, the specimens turn out to be problem-free and very productive. Meanwhile, various varieties were depicted - even in one brood.

SD and colouration: Their deep-bodied appearance, the relatively steep falling forehead and the hinted nose hump give the adult males of this species a hefty look. In its upper part as well as in the unpaired fins, the silvery whitish body is covered with bright ocher tones and lets us suppose forceful but also relatively problem-free Dwarf Cichlids. The pattern of the caudal fin is typical for members of this genus and is only seldom shown (shock, stress) in full colour tones, but rather only hinted, whereby the horizontal band of the males appears more as a wide zig-zag band and that of the yellow females in the breeding time is a line of coarse points. Intense blue colours can be seen in the males' cheeks and temples behind the padded lips in the form of patterns of spots and lines. A blotch pattern of fine red spots, of which the popular name was derived, lies between the blue lines. In both sexes, the caudal fin is truncated.

Total length: Male 8 cm, female 5 cm.

Maintenance in the aquarium and reproduction: See *A. macmasteri*.

S03935-3 *Apistogramma* spec. ›Rotpunkt‹, adult wild male.

Apistogramma spec. ›Segelflossen‹
A variant of *Apistogramma uaupesi* (see there).

Apistogramma spec. ›Smaragd‹
New scientific name:
Author and year:
Distribution: RÖMER was the first author to present this species in 1991 (b). He received the specimens together with some specimens identified by him as *A. regani*, from the collectors E. FRECH, A. WERNER and W. FRIEDRICH. As approximate location, they named a clear water stream in the area northwest of the city of Santarém. This is the area around Óbidos where the Rio Trombetas flows into the Amazon (cp. water sample in *A. geisleri* and map page 93).
SD and colouration: In this concern, RÖMER remarks: »The species is characterized by a clearly banded caudal fin, a conspicuously big caudal blotch and clearly pronounced bars on the body. Morphologically, the novices differ a bit from the *regani* type. Also the anal blotch of *A. regani* is absent. On the other hand, on the base of the pectoral fin there is an intense yellow orange blotch, as it is typical for species of the *eunotus* group. Besides the v-shaped bent seventh vertical bar on the caudal peduncle, the most conspicuous trait of these specimens are three to four stripes on the lower part of the body, as they were hitherto only known from *Apistogramma cruzi* KULLANDER, 1986. In an aggressive mood, the horizontal band is dissolved in several rectangular blotches and the described horizontal bands on the lower part of the body are more pronounced. When relaxed, the body of well developed males is covered by an intense blue green glimmer, so that I temporarily call the species "Smaragd-Apistogramma" (emerald *Apistogramma*).«

In the breeding season, female specimens are yellow to yellow brown, but in this colouration there are no pure tones: somehow it appears ›washy‹. A cheek stripe is visible and the horizontal band occasionally consists only of a pattern of five to six beaded blotches. The anterior shorter spine of the dorsal fin is deep black, on the other hand the black colouration on the anterior edge of the central fins is less intense.
Total length: Male 5 cm, female 4 cm.
Maintenance: RÖMER was the first to reproduce these specimens. »The first lay consisted of 220 eggs. After nine days, there were 132 young.

S04005-3 *Apistogramma* spec. ›Smaragd‹, adult offspring male.

Also later broods were comparatively prolific. After about three months, the young fish reached a length of three to four centimeters. Already at this size, they are able to reproduce,« the mentioned author remarks.

In the offspring the colouration of the young fish constantly improves with their further development and in further generations, so that now the ends of the dorsal and anal fins are more extended than in the wild specimens. Moreover, there are beautiful bright blue patterns on the gill covers which - at least in some specimens - are combined with red blotches.

Apistogramma spec. ›Tucuruí‹

New scientific name:

Author and year:

Distribution: The town of Tucuruí is near the hydroelectric power station with the same name, below the barrage in the drainage of the huge Rio Tocantins, one of the biggest clear water rivers flowing into the delta arm (Rio Pará) of the lower Amazon, just south-west of Belém. On its way to its mouth near the delta island Marajó, the Rio Tocantins has united with the equally big Rio Araguaia and therefore transports much water in its lower part. The hitherto introduced

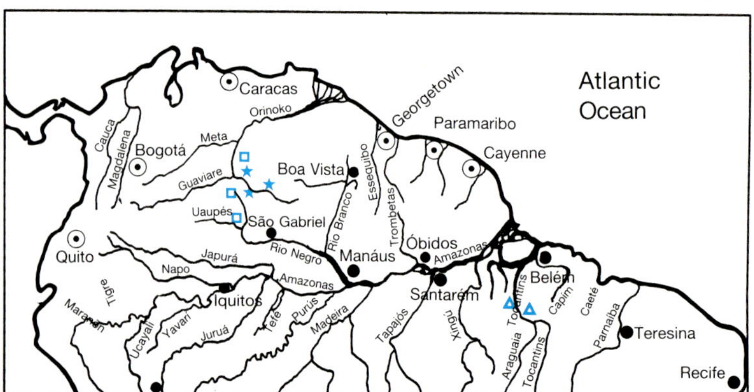

★ *A. spec. „Rotpunkt"*
□ *A. spec. „Vierstreifen"*
△ *A. spec. „Tucuruí"*

Biotope: Rio Tocantins,
in the main river, far upstream Cametá (Tassaisaú) - downstream the dam
Date: 25.6.1986
Time: 9.30 o'clock
Water movement: Still, but not sluggishly flowing
Water depth: Not measured
Water colour: Moderately clear

Laboratory results:

pH-value	6.60
Electrical conductivity (25°C), μS/cm	40
Acid capacity up to pH 8.2 (p-value), mmol/l	n.p.
Acid capacity up to pH 4.3 (m-value), mmol/l	0.35
As carbonate hardness, °d	1.0
Total Alkalis (Ca + Mg), mmol/l	0.14
As total hardness, °d	0.8
Calcium(Ca^{2+}), mg/l	3.6
Magnesium (Mg^{2+}), mg/l	1.2

Total Iron ($Fe^{2+}/^{3+}$), mg/l	1.28
Manganese (Mn^{2+}), mg/l	0.17
Ammonium ($NH4^+$), mg/l	0.08
Nitrite (NO_2^-), mg/l	0.09
Nitrate (NO_3^-), mg/l	2
Chloride (Cl^-), mg/l	4
Sulphate (SO_4^{2-}), mg/l	11
Total Phosphate (PO_4^{3-}), mg/l	0.11
Hydrogen Carbonate (HCO_3^-), mg/l	21
Carbonate (CO_3^{2-}), mg/l	n.p.
Oxidation Potassium Permanganate $KMnO_4$ consume, mg/l	16

n.p. = not provable

(and reproduced) specimens of this still scientifically undescribed species come from the surroundings of Jacunda Nova, where near the town of Tucuruí in combination with the mentioned power station a reservoir was built (cp. map page 134).

Habitat: The bottom area is mostly sandy, partially covered by debris, foliage and wood. The water figures can be seen in the table.

SD and colouration: This Dwarf Cichlid is different in the pattern of the head and body from other *Apistogramma* species: On a sand coloured background, specimens of both sexes show a pattern of fine brown short horizontal stripes. Between the eyes, above the padded lips a pattern of brown dots can be seen. In male specimens, the caudal fin which is sand coloured above its base, shimmers in its upper part as blue as the entire anal fin.

At the ends of the soft parts, both are well extended and pointed. The transparent caudal fin is truncated, to slightly rounded. Males and females show the cheek stripe continuously, whereas the horizontal band, the dorsal and caudal blotches are only shown depending on the mood (affection, stress).

The sandy colouration of the females is intensified during the courtship and breeding time.

Total length: Male 5 cm, female a bit smaller.

S04015-4 *Apistogramma* spec. ›Tucuruí‹, adult offspring male.

Maintenance in the aquarium and reproduction: The volume of the aquarium for these fish shall not be too little. Indeed, around themselves the males accept several females, but towards other males they are extraordinarily intolerant. Besides this, maintenance as well as breeding, in soft slightly acidic water does not cause any problems. Reproduction was successful at different places, whereby the female directly cares for the offspring, while the male cares for the active defence of the territory (male-mother-family). The fertilization of the eggs does not necessarily happen through the male swimming through the narrow entry into the cave. Moreover, it sets free its sperm in front of the cave's entry and pushes it into cave by subsequent fin movements. At a water temperature of about 27°C, the lay of roughly 100 eggs, develops in about 72 hours and after four more days, the development to the swimming young fish is completed.

Apistogramma spec. ›Vierstreifen‹

New scientific name:

Author and year:

Distribution: These specimens come from an area at the central Rio Orinoco with the centre of Puerto Ayacucho. As reported by RÖMER (1995), they are also present in the drainage of the Rio Uaupés (northwest of Brazil) and hence should inhabit the entire enclosed area (cp. map page 134).

Habitat: The water in the area of the central Rio Orinoco is extremely soft and acidic. Partly there is still an opulent rainforest. The bottom is covered with foliage and dead wood.

SD and colouration: Specimens of this new and scientifically still undescribed species can be called deep-bodied. In specimens of both sexes, the caudal fin is rounded. The dorsal fin is relatively high and not much extended, whereas the ventral fins are extremely extended. As their closest relatives *A. uaupesi*, *A. pertensis*, *A. meinkeni* and *A. iniridae*, the specimens impress rather less by their intense colours, than by their elegant body and the splendid finnage. Together with the mentioned species, they are combined in the *pertensis* group.

As basic colouration of the males, a grey green can be named, but the ventral part remains whitish. This part is covered by four rows of dark dots (popular name). The very fine cheek stripe is not very conspicuous and the vertical oval caudal blotch is separated from the body by a silvery stripe. The rounded caudal base shimmers bluish and shows five vertical rows of red brown spots. The anterior part of the anal fin shows bright blue tones, whereas the posterior soft part is covered with a pattern of orange dots. A horizontal band is clearly present. It runs from the eye to the mentioned silvery stripe in front of the caudal blotch.

In state of affection, there are six vertical bars. In combination with dark horizontal band which in this state now is also darker, the impression of a cheque pattern is shown. In its base, the high dorsal fin is brown red. Its entire edge is seamed silvery green. The exceptionally long ventral fins show a bright blue tone, their tips are coloured bright yellow. Dorsal and anal fin of the females are developed clearly less.

Total length: Male 8 cm, female 5 cm.

Remark: In 1996, specimens which could belong to this species were also occasionally imported from Perú together with other fish. Their exact origin could not be found out.

Maintenance in the aquarium and reproduction: As mentioned above, this new species is not widely distributed in the hobby. Its reproduction can be considered difficult, but at a conductivity of 150 µS and a pH of 5.5 it should be possible. Meanwhile, German offspring is offered in smaller numbers.

S03862-3 *Apistogramma* spec. ›Vierstreifen‹, adult wild female.

S03862-3 *Apistogramma* spec. ›Vierstreifen‹, adult wild male.

Apistogramma spec. › Weißsaum‹

New scientific name:

Author and year:

Remark: In 1991, the popular name for this by then still unknown species was introduced by WINDISCH. Already in 1985, similar specimens were presented as ›Balzfleck-Apistogramma‹ by KOSLOWSKI. But for example a difference between both is in the caudal fin which in the specimens of the latter species is transparent.

Distribution: The hitherto introduced dwarf-like specimens occasionally came from Puerto Inirida (east Colombia) together with *Dicrossus filamentosus*, there they have the native name ›Olga‹ (cp. map page 88).

Habitat: Similar to as given for *A. iniridae*.

SD and colouration: Males with a dorsal fin that in its central part is iridescent blue. The basis and the seam are black. The upper edge with the truncated membranes shows a colouration covered by white, temporarily orange tones. The same colours can also be seen as a pattern in the base of the dorsal fin as well as its soft part. The caudal fin with a pattern of black and orange vertical bars has a rounded to truncated posterior edge. Every scale has black edges so that the flanks appear reticulated. Above the centre of the flanks, there is an intensely iridescent blue green shimmer. The head is sooty dark and covered with several short, sky blue shimmering stripes.

In the females, the caudal fin is only slightly patterned and its posterior edge is truncated to rounded. The dorsal fin shimmers in turquoise and orange tones and every jutting out end of the spines has a fine white edge. In the more closely connected part of the soft dorsal fin, the white ends form a white seam (popular name). In the lower part, the yellowish flanks of the females are covered with a sooty tone, depending on the mood, but on the other hand in the upper part they remain purely yellow. The eyes (especially that of the female shown here) appear to be relatively big, which lets us assume a life in a dark environment (black water).

Total length: 3.5 to at the most 4 cm, whereby female specimens mostly are the (slightly) bigger and stronger ones.

Maintenance in the aquarium and reproduction: Here we refer to the information given by WINDISCH: »The courtship comes from the female as soon as she is ready and willing to spawn.

S04016-4 *Apistogramma* spec. ›Weißsaum‹, adult wild male.

Although in my tank with soft water and pH values around 4.5 the fish spawned, the eggs regularly decayed. Later, my friend from Rastatt/ Germany was able to make the larvae hatch. The pH value had been even lowered. He also told me that the dwarfs completely closed their breeding cave with fine sand. My specimens did not have that possibility and I do not exclude that this equally plays a role for the success or failure of reproduction. Due to the small size of the specimens - particularly of breeding females - they seem to have a high demand for shelter.«

S04016-4 *Apistogramma* spec. ›Weißsaum‹, adult wild female in breeding colouration.

Apistogramma staecki KOSLOWSKI, 1985

Meaning of the specific name (etymology): Dedication name in honour of the discoverer, author and breeder Dr Wolfgang STAECK from Berlin/Germany.

First description: Bonner zool. Beitr. 36 (1/2), pages 145-162.

Remark: The specimens were collected by the authors LINKE & STAECK (1992) in Bolivia - together with those of *A. linkei* (see there). Up to the scientific description, the specimens of this species (1984) were presented as ›Querstreifen-Apistogramma‹.

Synonyms: None.

Distribution: The authors LINKE & STAECK describe the type locality: It is situated in the southern surroundings of the city of Trinidad (road to El Colegio and Loreto) in the eastern lowland of Bolivia. This area of Llanos in the drainage of the Rio Mamoré, a big tributary of the lower Rio Guaporé which flows through the Rio Madeira into the Amazon, serves for cattle raising. During the rainy season, its neighbouring meadows are widely inundated. Ditches with a depth of 50 to 60 cm and a length of several kilometers, mostly covered by water hyacinths (*Eichhornia*) run besides the elevated road traces. Here the small Cichlids live in the areas that formed among the roots and the foliage. The free space between the roots of the plants on the one side and the more or less muddy bottom on the other side is inhabited by bigger, frequently predatory fish (cp. map page 106).

Habitat: The slightly troubled water (following LINKE & STAECK, 1992) has only a low total and carbonate hardness (<1°d) respectively a conductivity of 94 µS at a water temperature of 29°C (measured in the afternoon). The pH value is given as 5.7.

SD and colouration: A Dwarf Cichlid which until now for unknown reasons had problems finding popularity in the hobby. On a sand coloured background, the males show an intense pattern in the caudal fin which depends on the mood and during the courtship and breeding season is almost completely lost.

Only on the golden cheeks, then the stripe is a bit more intense. In its basal part, the dorsal fin is equally sandy coloured, but in the upper it shimmers bluish. Similar blue tones can be seen in the hard part of the anal fin. The posterior parts of the dorsal and anal fin with the soft fin rays which are not extended, also show the pattern of stripes and dots of the caudal fin. The caudal fin of the males is bilobated, in the females it is rounded. On a transparent to bluish basic colouration, in the males a pattern of brown dots can be seen which form vertical bands. During the courtship and breeding season, they turn into a red colouration which then also extends into the dorsal and anal fin. In elder males, the ventral fins are well extended and then they shine in red and blue tones and show white ends.

Total length: Male 5 cm, female 3 cm.

Maintenance in the aquarium and reproduction: Unfortunately, in the tanks of the hobbyists the representatives of this species did not remain a long time. Today, offspring specimens are hardly available.

In LINKE & STAECK (1992), the maintenance of specimens of this species is described as problem-free and the specimens are said not to have particular demands in water quality. Similar things are also true for reproduction, whereby it has to be considered that »a better development of the eggs and the larvae in softer acidic water« occurs. At a water temperature of 28°C, the larvae hatch after 55 hours after the fertilization. Then they are 3.6 millimeters long. After 160 more hours the young fish swim free, are taken around by the mother for their first ingestion and immediately consume *Artemia* nauplii as well as finest food for fry. »Despite repeated daily feeding and good water quality, the young fish grow only slowly. The species is not very productive«.

S04045-4 *Apistogramma staecki*, adult male.

Apistogramma steindachneri (REGAN, 1908)

Meaning of the specific name (etymology): Dedication name in honour of Dr Franz STEINDACHNER (1834-1919) from Austria, who following an invitation of Prof Dr Agassiz participated in a voyage from Boston around southern America to San Francisco from 1870 to 1874. Thereupon, in 1875 besides others he described *Apistogramma agassizii*. After his return to Austria, in 1876 he was appointed Director of the Zoologische Hofkabinett. In 1877, the appointment to the Director of the zoological collection of the ›k.k. Naturhistorisches Hofmuseum‹ (Royal and Imperal Museum of Natural History in Vienna). Due to his numerous merits, besides many honours he received the title ›Hofrat‹.

First description: Ann. Mag. Nat. Hist. (8) 1, pages 370-371.

Synonyms: *Heterogramma s.*; *Apistogramma ornatipinnis* AHL, 1936; *A. wickleri* MEINKEN, 1960.

Distribution: In Surinam (drainage of the Suriname and Corantijn Rivers) and Guyana (drainage of the Essequibo, Demerara and Mahaica River) as well as in the eastern (border) area of Venezuela in the Essequibo drainage (cp. map page 141).

Habitat: In various parts of waters like ponds, streams, ditches and (temporarily) flooded areas. Here the water is very soft, the pH value can be above or below the neutral point (7.0). Here most of the bottoms are also sandy or covered with vegetated rocks.

SD and colouration: Above the horizontal band which is shown depending on the mood, the body colouration of *A. steindachneri* is grey olive. Every scale has dark edges so that a reticulated pattern occurs. The ventral part remains whitish and is covered with a bluish shimmer. Vertically arranged dots cover the bright blue bilobated caudal fin whose lower edge can be yellow or intensely orange.

Below the eyes, there are worm-like bright blue markings. A lateral blotch is only temporarily visible. The cheek stripe as well as the caudal blotch are always - more or less intensely - visible. The long ventral fins are bright blue, their tips are yellow or orange red. Female specimens have a grey olive basic colouration, in

S04250-3 *Apistogramma steindachneri*, adult wild male.

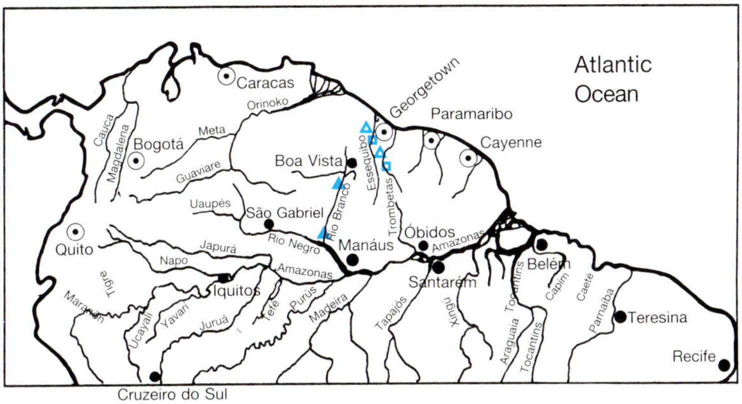

□ *A. ortmanni*
△ *A. steindachneri*
▲ *A. rupununi*

which the ventral part remains whitish. Females only have a truncated caudal fin, the dorsal and anal fin are clearly less extended than in male specimens. The not very much pronounced horizontal band is not always visible, whereas the lateral and the caudal blotch are always visible.

Total length: Male 9 cm, female a third smaller.

Maintenance in the aquarium and reproduction: Specimens of this species were reproduced in eastern Europe over a longer period of time and brought into trade in Germany. Maintenance and breeding do not cause remarkable problems to the hobbyist. For reproduction, the relatively big specimens shall not be kept in too small tanks. They lay up to 200 eggs.

Apistogramma sweglesi MEINKEN, 1961

Meaning of the specific name (etymology): In honour of the US American collector and breeder Kyle SWEGLES (›Rainbow Aquatic Nurseries‹).

First description: DATZ 14 (5), pages 135-139.

Synonyms: None.

Distribution: As origin of the type material (following information given by SWEGLES), MEINKEN names ›Letitia in Peru‹. Indeed the city of Leticia is in the border area of the three countries Colombia/Brazil/Perú, but it politically belongs to Colombia.

Habitat: The area around the city of Leticia at the Amazon is characterized by very soft, slightly acidic water which is absolutely free of nitrate (cp. water sample Rio Arara in *A. bitaeniata*).

Remark: The specimens that then were introduced by Swegles through Dr E. SCHMIDT-FOCKE from the former company ›Tropicarium Frankfurt‹ in Buchschlag, come from the mentioned area in the border area of the three countries rich of water.

As type material, two males with a total length of 73 respectively 67 millimeters and a female with 50 millimeters were at disposal. Following the explanation added to the first description, »after the publication of the examination result it passes over to the possession of the Zoologisches Staatsinstitut of Hamburg«.

Unfortunately, in this concern MEINKEN did not mention an identification number (= ZIMH). In his revision of the genus *Apistogramma* from 1980, KULLANDER adds that following his investigations (written comm. Prof Horst Wilkens, 1977) this material has to be considered undiscoverable.

Due to a written communication by MEINKEN (1976), all his *Apistogramma* samples shall have been given to the Senckenberg Museum of Frankfurt (SMF), but they are also not to be found here.

Considering the drawing added to the first descripton (?), by the habitus one is reminded of a Dwarf Cichlid (except the compact head), which today we know as *A. paucisquamis* KULLANDER & STAECK, 1988. But the specimens of this species were described from the central Rio Negro and are occasionally imported via Manáus.

Remark: Today *Apistogramma sweglesi* is generally considered synonym of *A. bitaeniata*. As the type specimens cannot be found any more, it should be difficult to ›let this species revive‹ nowadays.

S04270-3 *Apistogramma taeniata*, adult wild male.

Apistogramma taeniata (GÜNTHER, 1862)

Meaning of the specific name (etymology): Adjectival form of the Latin *taeniatus* = banded. Following the rules, here the specific name (ending "a") was adapted to the feminine gender of the generic name.

First description: Cat. Fish. Brit. Mus. 4, London, page 312.

Synonyms: *Mesops taeniatus*; *Geophagus* (*Mesops*) *t.*; *Biotodoma t.*; *Heterogramma t.*

Remark on the distribution: As collection point of the received type material (a specimen [?] with 42.1 millimeters [SL or TL?], Legat 1852) of the British natural scientist Henry Walter BATES the Rio Cupaí (**not** Rio Cupari!) was given. GÜNTHER adds the particular remark: »800 miles from the sea«.

In the same work (Vol. 4, pages 306 and 307), for *Crenicichla johanna* var. *vittata* or *C. acutirostris* the author remarks: »Rio Capím, Rio Cupaí, 800 miles from the sea«. In his Brazilian work (Vol. IX, 1954), FOWLER verifies this data, but corrects the river's name and correctly calls it ›Rio Cupaí‹. The Rio Capím flows into a delta arm of the Amazon near the city of Belém.

The Rio Cupari which is (as an alternative?) mentioned by KULLANDER, can not be equated with the Rio Cupaí, because this one flows into the Rio Tapajós some 150 km upstream the city of Santarém - only about 700 to 800 kilometers (not miles!) away from the sea. As *A. taeniata* is defined as the generic type, the whereabouts should be clarified; the addition of a letter is not enough (cp. map page 86).

In further works, for the members of this specimens a wide distribution was acknowledged:

Geophagus taeniatus STEINDACHNER, 1975: Óbidos, Tefé, Tabatinga (all Amazon respectively Solimões) and Rio Tapajós; *Biotodoma taeniatum* PELLEGRIN, 1903: Manáus and Tefé; *Heterogramma taeniatum* HASEMAN, 1911: Bragança (a city at the mouth of the Rio Caeté [Para], São Antônio de Guaporé, Santarém, São Luiz de Cáceres (Mato Grosso) and Manáus; BERTONI, 1914: Paraguay; A. DE MIRANDA RIBEIRO, 1918: Manáus.

Our impression: There is no other Dwarf Cichlid known with such a wide distribution. Probably due to the lack of knowledge of that time, representatives of other (similar) species (for example the present *A. caetei*) were included.

SD and colouration: Concerning the holotype, KULLANDER (1980) remarks: »The holotype and only known specimen is a very old and poorly preserved fish which has lost its colours and most of its scales. The fins are also damaged that their shape cannot be determined.«

Total length: The single available specimen has a length (SL or TL?) of 42.1 millimeters. FOWLER (1954, Vol. IX, page 277) made a drawing of the fish on the base of a photo by HASEMAN. Therefore, it has a moderately elongated shape, intense cheek and horizontal bands as well as a rounded caudal fin.

The caudal blotch is oval and vertical. The membranes of the soft parts of the dorsal and anal fin rays are well extended; the same is true for the ventral fins.

Maintenance in the aquarium and reproduction: A Dwarf Cichlid which indeed is known in the hobby, but nevertheless it seems to be kept very rarely. The species is regularly reproduced by some professional breeders at Zwickau/Germany in smaller numbers and offered to wholesalers. Maintenance and breeding as given for *A. hongsloi*.

Apistogramma trifasciata trifasciata

(EIGENMANN & KENNEDY, 1903)

Meaning of the specific name (etymology): Adjectival form of the Latin ›*tri*‹ = three and ›*fasciatus*‹ = striped: hence ›with three stripes‹. Following the rules, in this case the ending of the specific name ("a") was adapted to the feminine gender of the generic name. With this specific name, the author refers to the cheek stripe and horizontal band as well as to a third which (not always visibly) runs from the base of the pectoral fins to the anal part.

First description: Proc. Ac. Nat. Sci. Philadelphia 55 (2), page 536.

Synonyms: *Biotodoma t.*; *A. t. haraldschultzi* MEINKEN, 1960.

Distribution: The type material was collected by Dr A. BORELLI in the system of the upper Rio Paraguay (Arroyo Chagalalina). Later reports talk about collections in São Luiz de Cáceres, Villa Hayes, Campo Alegre, Bastos and Santo Antonia do Guaporé (HASEMAN, 1911). In our time, when interested hobbyists can themselves explore the habitats of the fish, distribution areas from the Rio Guaporé (Mato Grosso/Brazil re-

S04320-4 *Apistogramma trifasciata trifasciata*, adult wild male.

spectively in the eastern lowland of Bolivia) through the entire system of the Rio Paraguay and beyond its delta area into the Rio Paraná can be proved. The presently southern most distribution is in central Argentina, more exactly in the surroundings of the cities of Santa Fé (on the right bank, LINKE & STAECK, 1992) and Paraná (on the left bank) at the big river, directly in the delta area of the Rio Salado (cp. map below).

Habitat: The habitats can show quite different water qualities because (in the south) they partly pass through intensely urbanized or agriculturally used areas. Similarly, the water types (clear and/or white water) can contribute to changes of the water readings. Indeed, they are always found in water of very low hardness, but the pH value can be in the alcaline range (above the neutral point of 7.0).

One deciding to search for these fish, will almost always find them in areas with rich aquatic vegetation. When scanty, the overhanging vegetation on the banks has to serve as hiding places.

SD and colouration: The males of these Dwarf Cichlids remain relatively small and are moderately elongated and occur in varieties that have long and big fins as well as more modest ones. In this case (mostly in the aquarium), these are varieties ›created by maintenance‹: As can be seen often in the growth of the fish (for example in the Congo Tetra), males growing up in very soft water have longer membranes of the anterior dorsal and ventral fin rays, the end of the soft fin rays of the dorsal and anal fin are more extended and also the pink shimmering rounded caudal fin can grow higher.

The basic colouration of the body is silvery grey. A sky blue shimmer is on this colouration, which lets the throat and the ventral part remain white. The unpaired as well as the ventral fins show themselves in iridescent blue tones. On the forehead and the neck, there is an ocher yellow zone. Among the black markings, the cheek stripe and the horizontal band that slightly gets wider to the rear part of the body are almost always shown. A caudal blotch is absent.

Female specimens can be easily differentiated as they stay smaller than the males, show a yellow colouration on the head, body and fins (which in the courtship and breeding season is more intense) and during all the periods of time mentioned here they show an intense black colouration in the (normally extended) anterior dorsal and anal fin rays.

Total length: Male 6 cm, female 4 cm.

Maintenance in the aquarium and reproduction: The maintenance of these fish in the aquarium is relatively problem-free because their demands on the water quality are limited. Also the reproduction in the aquarium does not cause particular difficulties. Indeed, particularly for their productivity in reproduction, it turned out to be

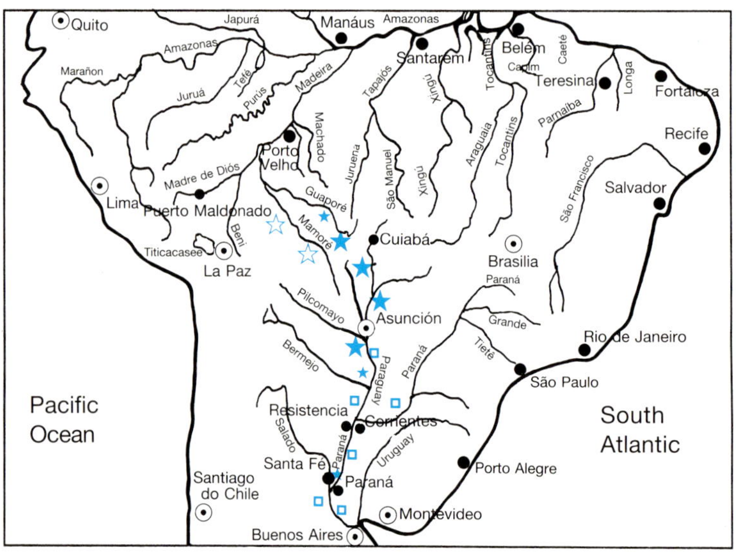

★ *A. trifasciata trifasciata*
□ *A. spec. „Paraguay II“*
☆ *A. trifasciata maciliens*

S04320-4 *Apistogramma trifasciata trifasciata*, adult wild female with a third band.

advantageous if the water is soft and its pH value is in the acidic range (below 7.0). The small females can produce 40 to 60 eggs.

Apistogramma trifasciata maciliensis

(HASEMAN, 1911)

Meaning of the specific name (etymology): After the owner of a plantation in the surroundings of the type locality, Mr B.A. MACIEL. Obviously, in the decisive part of the publication this name was misspelled. Then it should read "*macielensis*", as it was also used in a later part of the work. But due to the International rules, the misspelled name in the description of the formerly new subspecies is valid.

First description: Ann. Carnegie Mus. ser. 7, pages 360-361.

Synonyms: *Apistogramma trifasciatum haraldschultzi* MEINKEN, 1960, whose type material comes from the same type locality.

Remark: The specimens of this subspecies are those that meanwhile among hobbyists and breeders became known as ›*A.* spec. Mamoré‹. Examinations showed that contrary to *A. trifasciata*, in living specimens even depending on the mood the diagonal band running from the base of the pectoral fins to the anus is not visible; but in specimens preserved in alcohol, after a few days it is visible.

Distribution: As locality, the author mentions: »São Antonio de Guaporé, July 9, 1909. The variety is quite different from the typical Paraguayan form, but all stages of intergradation are present, and I have described it only as a variety in order to note the difference from the typical forms, which are also found in the Rio Guaporé« (cp. map page 144).

Habitat: The Rio Mamoré is a big tributary of the Rio Guaporé. Together with the Rio Beni, these three form the source rivers of the Rio Madeira. Additionally, the Rio Guaporé forms a section with a length of several hundreds of kilometers of the Brazilian/Bolivian border. The depression crossed by the big rivers and their tributaries, is widely swampy depending on the season of the year so that the fish living here have a possibility to extend their habitats.

The very soft and slightly brownish waters are relatively shallow and richly vegetated. Already early in the morning, during the southern summer the thermometer indicates a temperature between 23 and 25°C. There is opulent accompanying fauna. The high number ofTetra species is particularly conspicuous

SD and colouration: Male specimens are relatively deep bodied. A wide blue black horizontal band runs from the eye to the base of the caudal fin and ends with an equally blue black blotch. From the tip of the snout through the back, specimens of both sexes are olive brown. In the males, the mentioned caudal blotch is rounded

S03901-4 *Apistogramma trifasciata maciliensis*, wild male, blue.

by an ocher tone in its upper and lower part. Below the horizontal band, the specimens are bright ocher, the ventral part is whitish. The mouth is grey olive and below the yellow rounded eye, the head including the gill covers is golden yellow. Below the eye and behind it, the yellow area is covered with a sky blue line-formed marking. The pointed grey blue anal fin is extended almost up to the centre of the rounded caudal fin. Contrary to this, the ventral fins are only moderatly extended, showing a grey blue basic colouration and additionally are covered with a white blue anterior edge as well as with white blue tips. The form of the dorsal fin is similar to *A. t. trifasciata*, which means: The membranes of the first five dorsal fin rays are extended, whereby the membranes of the third and fourth spine are extremely extended. Above the base of the dorsal fin, an olive tone can be seen which in the central part of the fin rather turns into a bluish tone. The upper edge has a bright blue seam. The already mentioned membranes in the dorsal part that are extended either show a bluish tone, just like the caudal fin, or depending on the colour variety they are salmon red.

Until now, two colour varieties are known in which the extremely extended membranes of the dorsal fin and the rounded caudal fin show the tones mentioned before. In specimens of both varieties, a narrow dark cheek stripe is present. In state of affection, the dark horizontal band is less intense. It is then replaced by a washy blue zone.

In the female specimens, the membranes of the second and third dorsal fin ray are slightly extended. The smaller caudal fin is also rounded. An olive horizontal band runs from the eye to

S03901-4 *Apistogramma trifasciata maciliensis*, offspring male, red.

S03901-4 *Apistogramma trifasciata maciliensis*, female.

the base of the caudal fin. The equally olive dorsal part is darkened towards its upper edge. Below the horizontal band, the females are yellowish to whitish. The anal fin is bright blue. In courtship mood, body and fins of the female specimens are covered with an intense green yellow shimmer.

Total length: Male 8 cm, female 5 cm.

Maintenance in the aquarium and reproduction: Similar to that given in *A. t. trifasciata*.

Apistogramma uaupesi KULLANDER, 1980

Meaning of the specific name (etymology): To the origin of the type material, collected by the former Belgian King Léopold III together with the Belgian ichthyologist Dr Jean-Pierre GOSSE at the Rio Uaupés in 1967.

Remark: The type material (1 holotype [?] as well as the paratypes [28 ??, 15 CC and four of undefined sex]) came from the collection of the two discoverers mentioned above.

First description: Bonner zool. Monograph. (14), pages 122-126.

Synonyms: None.

Distribution: Although the type material comes from a small stream flowing into the right side of the Rio Uaupés (near Travão in the Brazilian federal state of Amazonas), this does not yet mean that the distribution of this species is limited to this narrow area. As far as could be discovered, *A. uaupesi* is identical with the species that are already kept in the aquaria of some hobbyists: the ›Rotkeil-Apistogramma‹ as well as the ›Segelflossen-Apistogramma‹. Both colour forms - the former with a reddish wedge pattern in the caudal fin - exist in the upper Rio Negro, even probably (sympatrically with *Dicrossus filamentosus*) in the affiliated Rio Orinoco. The variety from Brazil (?) shows the red pattern, the one existing up north in Colombia shows a more or less clear caudal fin.

This rarely imported representative of the genus is closely related to *A. pertensis* and *A. iniridae*. In this beautiful and elegant Dwarf Cichlid, the lyra-shaped caudal fin has particularly extended tips. Even among experts, the specimens are not widely spread. This is possibly caused not only by the fact that Cichlids of this species were imported only in a few specimens occasionally together with *Dicrossus filamentosus*. In their habitat, by the collectors mentioned above these fish were found together with *A. elizabethae*, *A. brevis*, *A. personata* and *A. meinkeni*. RÖMER (1995) reports also about an occurrence of *A.* spec. ›Breitbinden‹ as well as of *A.* spec. ›Vierstreifen‹ in these habitats, whose imports normally come to us from habitats in the upper Rio Orinoco and therefore from Colombia (cp. map page 61).

Due to statements by SCHMETTKAMP (MAYLAND pers. comm.; cp. also BORK, 1996), an affinity with the species known as ›Blutkehl-Apistogramma‹ can be excluded as during breeding attempts, the females were erroneously mixed and with the wrong partners there was no mating. After the error had been detected and the right partners were brought together, the following day spawning could already be observed.

Habitat: The water in the upper Rio Negro, as well as in the Rio Uaupés which flows into it, is extremely soft and clearly acidic (cp. water sample ›Upper Rio Negro‹ in *A. diplotaenia*). As here the rainforest is still present, one can assume that the sandy bottoms of the waters are covered with foliage and branches. RÖMER (1995) describes the habitat of representatives of the species as follows: »*A. uaupesi* can mostly be found breeding at steep banks with a faster water current, where dense roots of *Ficus* offer good hiding places. Larger groups of semi-adult specimens which can have up to one thousand specimens, regularly occur on sandy parts that are completely free of hiding places.«

SD and colouration: The common name ›Rotkeil-Apistogramma‹ for the probably more popular variety refers to a red, wedge-shaped colour zone on both sides of the black lateral band of the males that ends in the caudal fin. The caudal fin itself is elongated; its upper and lower edges are red brown, have white seams and two long filamentous extensions.

Already in 1980, the first specimens were described by KOSLOWSKI. He discovered the small Cichlids among imported specimens of *Dicrossus filamentosus* and *A. pertensis*. The specimens have an elongated body. Their males have a high dorsal fin which is sustained as a ›one-piece sail‹ by membranes that are united at their outer edges. The end of the soft part of the dorsal fin as well as that of the anal fin is only moderately extended. The anal shimmers intense blue. Patterns of blue points and lines can be seen on the cheeks, the gill covers and the throat. A snout stripe and a horizontal band run from the upper part of the mouth across the flanks and the caudal peduncle into the base of the caudal fin. A cheek stripe is present. A certain similarity to *A. pertensis* and *A. iniridae* is given; therefore all are placed in the *pertensis* group. The male representatives of the two mentioned species can indeed be recognized by the rounded caudal fin. Finding the suitable females can occasionally be a problem because just like those of the mentioned species they also have golden yellow courtship and breeding dress, have a shorter dorsal fin than the males and a slightly rounded respectively truncated caudal fin.

Total length: Male 9 cm, female 6 cm.

Maintenance in the aquarium and reproduction: The representatives of this species are characterized by a particularly aggressive behaviour. This aggression is obviously not only limited to the males of the same species. Already twice, I (Bork) had to make the rather painful discovery that the male killed the partner destined to him, already in one or two days - despite sufficient hiding places in a tank with a volume of 50 liters. Hereby, the possible reproduction of this interesting species is made more difficult additionally to the water conditions that are extremely difficult to achieve. To our knowledge, reproduction in the aquarium has not yet been possible. Despite good water conditions with a conductivity of about 100 µS and a pH value of 5.5, the eggs did not develop.

S04360-3 *Apistogramma uaupesi*, adult wild male during lateral threatening.

S04400-3 *Apistogramma urteagai*, adult wild male.

Apistogramma urteagai KULLANDER, 1980

Meaning of the specific name (etymology): Dedication name in honour of the Leader of the expedition, Jorge Andrés Urteaga Cavero, during which the type material was collected.

First description: Cichlid fishes of the Amazon River drainage of Peru. Swed. Mus. Nat. Hist., Stockholm (431 pp.), pages 163-168.

Synonyms: None.

Distribution: The holotype was caught in the system of the Rio Madre de Dios (Lago Tupac Amaru) near the town of Puerto Maldonado. Further material comes from a swampy area on the way to the Lago Sandoval, the system of the Rio Tambopata, the Rio La Torre as well as from the Laguna Cocococha in the Reserva Natural de Tambopata (cp. map page 86).

Habitat: The specimens exist in different kinds of habitats. Sandy bottoms are partly replaced by swampy and muddy bottoms. On the other hand, the waters are continuously very soft (total and carbonate hardness <1°d) and clearly acidic (pH 6.0). Very often there is emersed or submersed vegetation (*Echinodorus*). This does not exclude twigs and strong branches as well as foliage so that the fish which constantly search for shelter find hiding places. In zones with a less strong current, swimming plants offer more cover. Other small water gullies of the surrounding together with their pond-like extensions as well as rest water pools in the rainforest have dense vegetation. Here the specimens live side by side with their relative *A. luelingi*.

The lakes in the mentioned distribution areas (lagunas) of the flooding area of the Rio Madre de Dios are fed by white water rich in sediments, but with the time the sediments accumulate on the ground and are only stirred up by men entering the water. Among the further inhabitants of these waters are the omnipresent representatives of the genera *Crenicichla* and *Mesonauta* (here *M. festivus*). Others are members of the species *Aequidens tetramerus*, *Cichlasoma boliviense*, *Crenicara puntulatum* and *Satanoperca jurupari* - to name but the best known aquarium fishes among the Cichlids. STAECK (1993[c]) stated that the specimens show a high tolerance towards relatively low temperatures.

SD and colouration: In the aquarium, dominant males show a colouration like the specimen pictured here. Moreover, changes in colouration and pattern depending on the mood are possible.

Apart from the size, sexual differences can be seen in the simple grey beige basic colouration of the females which above the centre of the flanks show a black horizontal band. In their courtship and breeding dress, the females then show a golden yellow colouration including head and body. The mentioned horizontal band then can mostly be seen only as a line of spots.

Total length: Male 6 cm, female a third smaller.

Maintenance in the aquarium and reproduction: This relatively problem-free fish prefers slightly acidic water where it is even more productive! Also because of their conditional peacefulness, they are quite suitable for a correspondingly established community tank. Compared to some other relatives of the genus, in attempts to keep their females in a ›harem‹ with only one male, the specimens turned out to be rela-tively monogamous. Even as the specimens of this species do not show considerable colour attractiveness, but they are recommended for the beginner in Dwarf Cichlids particularly because of their interesting behaviour.

Apistogramma viejita KULLANDER, 1979

Meaning of the specific name (etymology): Diminutive of the Spanish ›vieja‹ = old = ›little old one‹. A name taken from the Colombian home of the fish which here - related to the breeding behaviour of the Dwarf Cichlids - means as much as ›little mother‹.

First description: Zool scripta 8, pages 71, 73-74.

Synonyms: None.

Distribution: The type material was composed after a collection trip by HONGSLO. These preserved specimens (= colour form I) came from a stream sending its water through the Rio Yucao (mouthing north of Puerto Gaitan) into the Rio Meta in East Colombia. Mainly proved by collections of aquarists, today we know a distribution of the fish in the vast drainage of the Rio Meta, a left sided affluent of the Rio Orinoco, about from Villavicencio down to (the east of) Puerto Gaitan (cp. map page 88).

Habitat: Just like in the neighbouring Venezuela, the llanos in the ›Oriente‹ of Colombia serve for cattle raising. Vast pasture lands stretch which are only interrupted by natural or artificial waterholes (›Lagunas‹), that are often accompanied by palms and thus visible from greater distances. The better the conditions for the fish are, the higher is their population density. Of course, this requires sufficient hiding places for adult specimens as well as for the offspring of different size. Therefore, the density of the vegetation at the banks in the very soft water which thus is poor in nutrients and clearly acidic depends on the consistency of the ground. It can be of coarse sand, to muddy. The nutrients contained in the ground can support vegetation or simply let it exist in a puny way.

S04440-4 *Apistogramma viejita*, adult offspring female displaying breeding colouration.

S04440-4 *Apistogramma viejita*, adult offspring male (reproduction Hüser).

In this vast area and the mostly connected waters (even if only temporarily), besides the Dwarf Cichlids one can find many aquarium fishes, so to name the Cichlid *Aequidens metae*, the catfish *Corydoras metae*, *Geophagus daemon*, little swarm Characins like *Hemigrammus bleheri* (cp. MAYLAND: ›Das Gesellschaftsaquarium‹, same publisher), *H. metae*, *Hypessobrycon pulchripinnis*, *Carnegialla strigata* or *Megalamphodus sweglesi*, *Moenkhausia eigenmanni* and *M. metae*; finally during our collecting trips in our nets we also had Cichlids of the genus *Mesonauta* and young specimens with a length of 4 to 5 cm of the large species *Hoplarchus psittacus*.

SD and colouration: The close relationship with *Apistogramma macmasteri* can also be detected optically, this species exists in the same area (›Villavicencio-Apistogramma‹) and *A. viejita* is placed into the same (*macmasteri*) group.

Already in the first edition of their book, LINKE & STAECK (1984) presented three different colour forms, but until today only the colour form I (perhaps in artificially improved colouration?) could be maintained, whereas the particularly attractive colour form II (with blood red dots and lines in the white throat part, much red in the caudal fin and a reddish glimmer in the dorsal fin) to a large extent has vanished from the hobby. In contrast to this, the colour form III is quite poor in colours.

The females of *A. viejita* are quite uniform. The golden yellow colouration on the head, body and fins is covered with deep black markings. The cheek stripe runs into the anterior throat part and horizontal band which is dissolved in a row of rectangular blotches ends in front of the vertical oval caudal blotch. Occasionally, also dorsal blotches are shown. The first three or four dorsal fin rays are as black as those of the posterior ventral fins.

Total length: Male 7 cm, female 4 cm.

Maintenance in the aquarium and reproduction: The specimens in the trade are mostly offspring For reproduction the water should be very soft and poor in nitrate, just like in the habitats where indeed, the pH value is mostly also in the clearly acidic range, but this does not seem to be too important for the success of reproduction - as long as the ph-value is below the neutral point. Otherwise, breeding occurs as given for *Apistogramma macmasteri*.

The genus *Apistogrammoides*

The only representative of this genus comes from the area around the Peruvian city of Pucallpa at the Rio Ucayali. When the US American editor (›The Aquarium‹) Albert J. KLEE made the type material available to the author (MEINKEN), he was thinking about having in front of him just another represen-tative of the genus *Apistogramma*. But MEINKEN soon found out that the specimens indeed have many similarities with those of *Apistogramma*, but they differ in a number of traits. The generic name with the ending ›*-oides*‹ (= similar) shall refer to the close relationship.

The discriminating traits are in the higher number of anal fin spines (8 instead of the common 3, formula A VIII/5-6) as well as in the position of the anterior lateral line. Contrary to the genus *Apistogramma* (here the anterior lateral line is only a half a scale width away) in *Apistogramma* the distance is always more than one scale width. KUL-LANDER (1980) places the genus together with *Apistogramma*, *Geophagus*, *Bioto-doma*, *Acarichthys* and others into the generic group of the *Geophaginii*.

Apistogrammoides pucallpaensis

MEINKEN, 1965

Meaning of the specific name (etymology): After the locality, a stream in the surroundings of the city of Pucallpa at the Rio Ucayali in Perú, one of the so-called source rivers of the Amazon.

First description: Senckenbergiana biol. 46 (1), pages 47-53.

Remark: For the scientific description, the author had four preserved specimens to his disposition. In extracts, KULLANDER (1986) comments MEINKEN's work from 1965 as follows: (1.) MEINKEN's count of the anal fin rays, A. VIII.5 in three and VIII.6 in one of the type specimens is not correct, as three specimens (including the holotype) have A. VII.6 and one specimen has VIII.5. (2.) Contrary to MEINKEN's assertion the upper gill arch (epibranchial) has one lobe - whereas it is smaller. (3.) The gill rakers are small, not well developed and despite their number appears uncertain (at least two to three), it is substantially lower than MEINKEN's ›more than a dozen‹. (4.) The teeth in the mouth, as sketched by MEINKEN, obviously include wart-like tissue. Most of the other described data are equally unprovable, but those listed here are important because they are able to form a genus and partly indicate that *Apistogrammoides* has to be considered a transition form from *Apistogramma* to *Cichlasoma*.

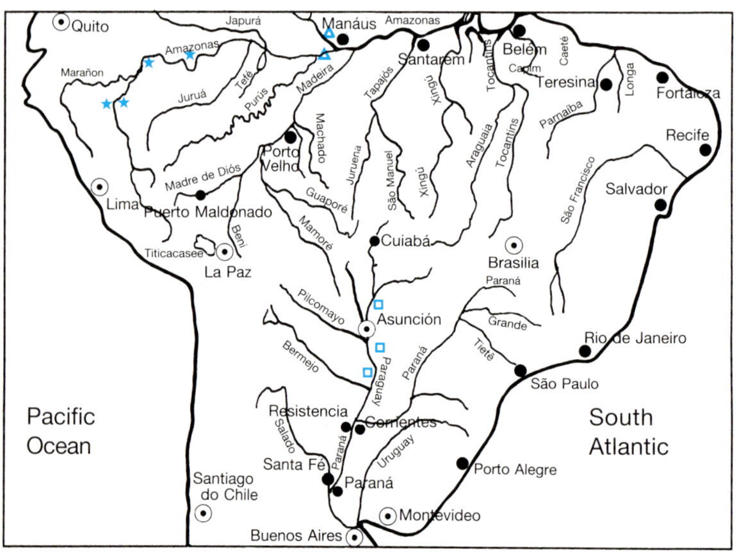

★ *A. Apistogrammoides pucallpaensis*
□ *A. spec. „Paraguay I"*
△ *A. spec. „Gelbwangen"*

S04500-4 *Apistogramma pucallpaensis*, adult offspring male.

Synonyms: None.

Distribution: The distribution area of this species is proved for the surroundings of the town of Pucallpa in Perú (Rio Utiquinia, lagoon near Paca Cocha, La Cabaña near Yarina Cocha) through the drainage of the Rio Pacaya (province of Loreto/Perú) into the drainage of the upper Amazon in Perú (Sacarita del Tuyé on the other side of the town of Pebas) up to Colombia upstream the city of Leticia (pool on the island of Santa Sofía) (cp. map page 154).

Habitat: In its readings, the water in this vast distribution range is irregular and at the upper Amazon, it can be from ›extremely soft‹ to ›soft‹ (total and carbonate hardness <1 °d, conductivity 120 to 280 µS). On the other side, in the areas at the Rio Ucayali with its tributaries from the Andes where the species was found, the readings are (locally) higher: Here the water can show readings from ›hard‹ to ›quite hard‹. In any case, the pH value is in the ›neutral‹ range (= 7.0) or slightly above it. The specimens are preferably found in quiet zones near submersed as well swimming plants.

The area around the 17 km long and horseshoe-shaped Lago Yarina Cocha near the city of Pucallpa at the Rio Ucayali is considered the type locality of *A. pucallpaensis* (name). In the recent decades, the lake was frequently examined. Its water is as black as coffee. The richness in fish is overwhelming. As the most conspicuous swimming plant, *Pistia stratiotes* is to be named which covers wide parts of the lake. From a zoogeographical point view, it is remarkable that here at the eastern slopes of the Andes and in the centre of the fresh water area of the Amazon, many fish can be found whose relatives come from the sea and which thereby support the thesis that in this area the sea had entry and respectively exit to the Amazon lowlands. This was the case in the Meozen, which means 10 to 20 million years ago. Fish like the horn pike *Potamorrhaphis guianensis*, *Colomesus psittacus*, *Achirus achirus* which is similar to a hand-sized plaice or the Amazonian freshwater sardine *Anchovella jamesii* prove this.

For the widely extend distribution range (from about Pucallpa up to Leticia), water samples taken by various authors (Lüling, 1980; Staeck, 1986[c]) show not only differences caused by the different seasons of the year, but also by geography. But even for the area around the city

S04500-4 *Apistogramma pucallpaensis*, adult wild female with eggs.

of Pucallpa, in the same year STAECK (1987) found rest water ponds near the Yarina Cocha with 26°C and 279 µS (9°dGH, 9°dKH, pH 7.1) and some near the Paca Cocha (Pucallpa) with 23.1°C and 510 µS (17°dGH, 16°dKH, pH 7.1). When it comes to water hardness, these readings are clearly differing from each other.

SD and colouration: *A. pucallpaensis* is a dwarf among the Dwarf Cichlids. In their normal colouration, the specimens appear plain. Only in breeding dress, do the males increase in attractiveness, their head turns golden yellow, the membranes of the unpaired fins as well as those of the ventral fins take a bordeaux red colouration and only the body keeps its olive green basic colouration. In the black pattern, only the horizontal line is conspicuous, it ends in a mostly triangular vertical caudal blotch (with one corner pointing towards the head).

The courtship and breeding colouration of the females is lemon yellow. Here the black markings are more conspicuous and besides forehead and cheek stripes, the black anterior dorsal and anal fin rays catch our attention.

Total length: Male 5 cm, female 3.5 cm.

Maintenance in the aquarium and reproduction: Breeding tanks should not be too small because among each other, they can become quite aggressive. Breeding females can attack the disturbing males that much so that these finally suffer from the injuries.

For maintenance and breeding the water does not have to have the extreme readings as known from the Dwarf Cichlids of central Amazonia. In order to ensure the necessary shelter for the small Cichlids, a rich vegetation in the aquarium reaching up to the surface is recommended. The installation of sufficient hiding places, among those caves, is obligatory. In mating, it turns out that polygamy is not so pronounced as found in the genus *Apistogramma*. Following the observations of our own breeding attempts and reports of other breeders, the male only mates with one female. The eggs are deposited by the female near the selected hiding place. In this case, unlike in most females of the genus *Apistogramma*, it does not have to be the ceiling of a cave. It can happen that the eggs are deposited at a vertical wall as well as on the bottom of a cave or directly near a cave in a gap of a root or a stone. The lay can comprise of up to 80 whitish eggs.

The female cares for the brood alone. After about three days, the larvae hatch and in four to six more days they develop into free swimming young fish. Now as the young fish have to care for food independently, the father is also accepted for guidance. The first feeding (as usual) happens with *Artemia* nauplii which are immediately consumed by the young fish. As soon as the young fish have survived the first days, they grow fast.

The genus *Biotoecus*

In the year 1975, in his work ›Beiträge zur Kenntnis der Chromiden des Amazonenstroms‹ (Contribution to the knowledge of the Chromids of the Amazon River) STEINDACHNER created the genus *Saraca* and defined the species *S. opercularis* described there as generic type. Later it turned out that this uncommon name was already used by an entomologic work by WALKER (1865) and was thus preoccupied. But it took 28 years until EIGENMANN & KENNEDY reached a conclusion and as an alternative, in 1903 they introduced the genus *Biotoecus*. It remained monotypic up to KULLANDER's description in 1989.

The meaning of the generic name derives from Greek and means as much as *biotos* = life and *oikos* = house, home. Probably the authors assumed that the affiliated species is a mouthbrooder (not a cave breeder, as we know today).

Biotoecus dicentrarchus KULLANDER, 1989

Meaning of the specific name (etymology): ›With (only) two spines (hard fin rays) in the anal fin‹.

First description: Journ. Nat. Hist. 23 (1), pages 225-260.

Synonyms: None.

Distribution: In the drainage of the central Orinoco, in the area between the cities of Puerto Carreño (Colombia) and Puerto Ayacucho (Venezuela) in the north up to the mouth of Rio Guavare (with the lower Rio Inirida) near Puerto Inirida (Colombia) in the south.

Habitat: In this area, the water is extremely soft and clearly acidic. Near the banks, the bottoms of the waters that are partly interspersed with rocks are covered with foliage and branches (›dead wood‹). In small more narrow waters, the vegetation at the banks often grows into the water, gives shadow and additionally shelter to the fish.

SD and colouration: Contrary to most of the other Cichlids whose anal fin has at least three hard rays, the representatives of this species have but two (name). Following information given by the author, the specimens of this species shall remind a bit of those of *Dicrossus filamentosus* with which - at least in some habitats - they exist ›sympatrically‹. Another discriminating trait compared to *B. opercularis*: the caudal peduncle is shorter, so that the specimens appear more compact. The membranes of the anterior two hard fin rays in the dorsal fin are extended.

Total length: Following the first description, the specimens should only be about 4 cm long.

Maintenance in the aquarium and reproduction: Similar to that described in *B. opercularis*: cave breeder.

Biotoecus opercularis (STEINDACHNER, 1875)

Meaning of the specific name (etymology): From *operculum* = gill cover; here referring to the dark blotch (which depending on the mood is only slightly visible): ›With a particular trait on the gill covers‹.

First description: Sitzber. kais. Ak. Wiss., Mathem.-Naturwiss. Klasse 71, pages 125-127.

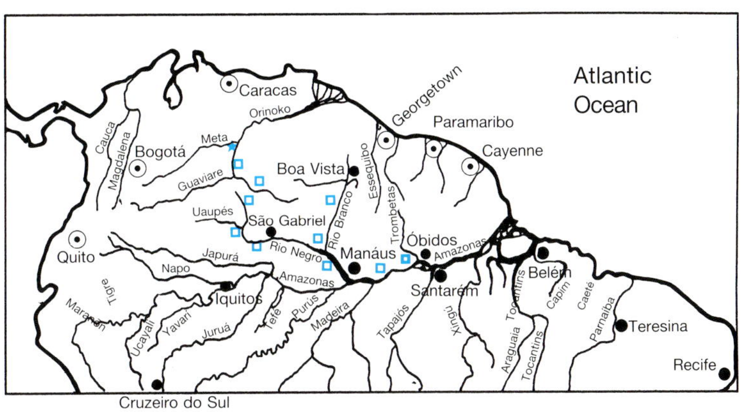

★ *Biotoecus dicentrarchus*
□ *Biotoecus opercularis*

Synonyms: *Saraca o.*; *Biotaecus o.* JORDAN, 1917-1920.

Distribution: The Lago Saracá which is named as type locality is formed on the left side of the Rio Urubú (after the confluence with the Rio Anebá and the Rio Caru), about 20 km west of the island and town of Silves. In 1985, BLEHER and MAYLAND discovered the fish in a black water lagoon a few kilometers east of the island in the delta of the Rio Sanabani. They were given to a renowned breeder for reproduction, but unfortunately the hoped success with these specimens did not occur at the time.

Today we know that the specimens of this species are widely spread in northern Brazil and even localities in Venezuela were known (cp. map). Although from its distribution range constantly considerable lots of various ornamental fish species were exported to the USA and to Europe, before the described rediscovery specimens of *B. opercularis* were never discovered occasionally together with imports of other fish. As further locations are known, also specimens of this species came to us. Meanwhile, they are regularly reproduced by various breeders.

Habitat: The widely distributed habitats may lead to differing results of measurements of water samples, but here I can refer to the samples given for the area at the Rio Urubú (cp. page 155).

SD and colouration: *Biotoecus opercularis* has a pointed head with a front pointing mouth and big eyes. The slender body is elongated in a wedge-like way and ends in a conspicuously thin caudal peduncle. The equally elongated dorsal fin is clearly split in a smaller part with hard fin rays (spines) and a bigger part with soft fin rays. The seven hard rays have the same length and are coarsely sawn. In the posterior part, the soft fin rays become noticeably longer and are very finely sawn. In elder males, the posterior part of the dorsal fin has a fine filament. The anal fin is long and pointedly extended, whereas the strong bent ventral fins only have a short extension. In elder males, the upper as well as the lower edge of the moderately high concave caudal fin can have long filaments. But normally, only the upper filament is fully developed. With their way of life which is very much bound to the bottom, they often ›sit‹ on their ventral

Black water lagoon in the drainage of the Rio Sanabani near Silves: capturing *Biotoecus opercularis*

S07320-3 *Biotoecus opercularis*, pair in neutral mood.

fins and the caudal fin. Above all, in maintenance on fine, sharp quartz sand the tender membranes of the fins are really polished by this behaviour. Therefore, the lower lobe of the caudal mostly appears to be clearly rounded: This is why the formation of the lower filament is only possible on a very soft and smooth bottom.

The females can be clearly distinguished from the males by their essentially smaller dorsal fin as well as by the smaller and less pointedly extended anal fin. The dorsal fin of female specimens mostly only has five hard fin rays; they are equally sawn more coarsly than the essentially bigger soft part. The dorsal and caudal fin of the females do not have extensions.

The colouration of the females is tender and thus less conspicuous. In normal mood, they are coloured grey green or grey blue. The ventral part is coloured whitish. On the other side, the lower part of the head shows tender yellow to golden tones. On the back, from the beginning of the dorsal fin to the caudal peduncle five dorsal blotches are visible. Near the lateral line, above all in the posterior part, temporarily an unclear blotch pattern appears. In the centre of the body, a lateral blotch is present which in female specimens is mostly more pronounced.

The short bent ventral fins show a tender green yellow colouration. On the other hand, the rest of the fins appears transparent and more or less uncoloured. In the males, the edges of the dorsal and anal fin, including the filaments, have lilac red seams. In courtship mood, the blotch pattern is less intense, the males appear silvery with slightly glittering pastel colours in lilac, red and green (photo). On the body of the females, the lateral blotch is now particularly pronounced.

On a dark bottom, the tender pastel colours can be seen much better than on bright quartz sand.

Total length: Male 5 cm, female a bit smaller.

Maintenance in the aquarium: For keeping these fish, aquaria with a length of 50 cm are enough. For well-being, fine material on the bottom is very important, this means: If possible, no gravel is to be used (it practically does not exist on the bottoms of the waters in Amazonia!). As the specimens can be very shy, when installing the aquarium the aquarist should include a sufficient number of hiding places on the bottom.

At temperatures around 28°C and water readings around 150 µS as well as a pH value of 6.5 the fish are feeling very fine and - if sexually mature - they will soon start reproducing. Live food and *Artemia* nauplii are the preferred diet, but also worms are gladly accepted. If *B. opercularis* is socialized with peaceful small Characins under the mentioned conditions, these normally shy Dwarf Cichlids will more likely show their interesting behaviour.

Breeding: This species is a pronounced cave breeder! The amber coloured eggs are deposited at the ceiling of the cave. For efficient reproduc-

S07320-3 *Biotoecus opercularis*, offspring male in courtship colouration.

tion, the use of plastic tubes with a diameter of about 2 cm proved best as an offered spawning cave. The tubes with a length of about 10 cm are simply halfway pressed into the quartz sand. The final installation of the breeding cave is subsequently done by the fish themselves. The entry of the cave is closed up to the desired height by shovelling the quartz sand. This procedure is initiated by the male and can obviously be understood as mating ritual. Hereby the mouth is pushed forward and under a declination of about 30 degrees, the sand is pushed towards the entry of the cave. Soon also the female will participate, whereby it more frequently interrupts this work in order to spew sand into the direction of the entry. If the offered plastic tube

S07320-3 *Biotoecus opercularis*, offspring female in courtship colouration.

is too high above the environment, considerable masses of sand have to be moved. By this a real breeding hill is built.

After the spawning, the females takes over the care for the eggs. After about three days the larvae hatch and then they hang wriggling at the ceiling of the cave. In this stadium, the female occasionally leaves the cave, but it mostly stays in direct closeness of the entry, whereas the partner takes care of the defence of the outside territory.

But if the male comes too close to the cave, it is driven away vigorously by the aggressive fe-male. If in a relatively small tank the necessary shelter is absent, the male has problems with-drawing from the attacks of the female. In order to exclude injuries, then the endangered male has to be taken out of the aquarium. The breed-ing instinct of most of the females is very much pronounced so that even after massive inter-ventions in the breeding aquarium, the female subsequently immediately continues taking care of the fry.

After 8 days, the young fish swim free and are lead by the female. The young are still very small and therefore they cannot yet eat *Artemia* larvae as first food. But during the first week feeding Liquifry stood the best. Then *Artemia* nauplii can be given.

The small grow relatively slow. At the age of about seven months they are sexually mature.

The genus *Crenicara*

In 1875, STEINDACHNER created the genera *Crenicara* and *Dicrossus*. By his famous revision of the South American Cichlids (1905/1906), REGAN comprised the species of both genera in *Crenicara*. As this one was the first to be mentioned, he gave it priority (but unfortunately he erroneously used the name ›*Cren**a**cara*‹ which following the rules is invalid), so that the name *Dicrossus* fell back into synonymity. Due to their pattern of rectangular blotches, in German the representatives of both genera are called ›Schachbrettcichliden‹ (Chess Board Cichlids).

But nowadays, many scientists start to put the species that in former times were com-prised only in one genus into generic groups or also in different genera (newly created or revalidized ones). By this for example, the family relationships and the coherences in phylogeny can be better explained and defined. KULLANDER took the occasion of his creation of the genus *Mazarunia* (1990, see there) in order to revalidize the genus *Dicrossus* that was ›shut down‹ since the beginning of this century and therefore to place the two smaller and considerably more slender species *D.*

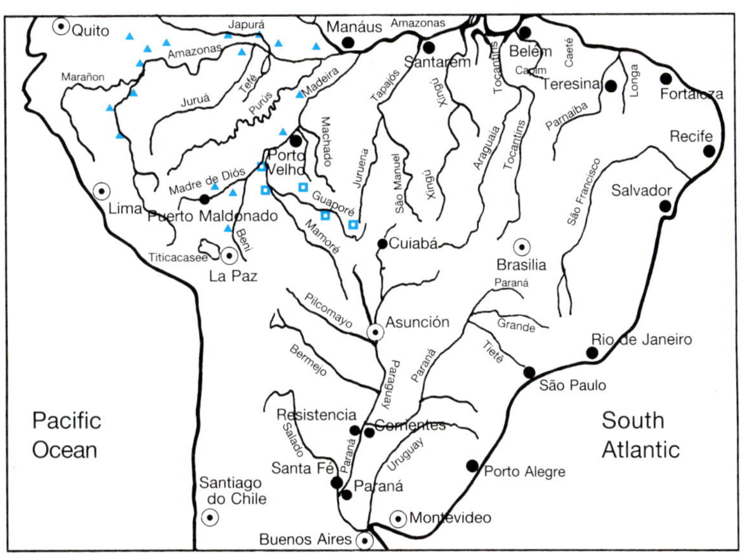

□ *Crenicara lutruncularium*
▲ *C. punctulatum*

filamentosus and *D. maculatus* into their original genus. Therefore, the deep bodied and essentially larger growing specimens of *C. punctulatum* would remain in the genus as the only species, if not in 1990 with *C. latruncularium* KULLANDER & STAECK would have described another *Crenicara* species.

In South America, the representatives of the genus have a wide distribution reaching from Guyana in the north-east through the Amazon and the Rio Madeira into the upper Rio Guaporé in Brazil and the upper Rio Beni in Bolivia (cp. map pp 159 and 164).

The members of the two mentioned and meanwhile again separated genera have several optical traits in common: A chess board like pattern of rectangular blotches on the body flanks, a conspicuously small mouth and red ventral fins in the female specimens which - at least in females of the genus *Dicrossus* - only show up after the first spawing.

Specimens of the genus *Crenicara* (due to lower demand?) are imported relatively rarely. Also in their natural habitat, they are never found in greater assemblages so that catching them turns out to be more difficult. In the aquarium, they should be offered an opulent vegetation, thereto a few shy accompanying fish. Otherwise, representatives of *Crenicara* kept alone, which means in a tank with only one species, can live in a very withdrawn way so that the hobbyist seldom sees them in the front part of the aquarium.

Contrary to the previously mentioned *Apistogramma* species, the members of this genus are no cave breeders. The females rather deposit their eggs openly on a horizontal substratum, a leaf, a stone or a similar thing. After the fertilization of the eggs, the female alone takes care of the brood. The male is driven away from the closer vicinity and takes over the securing of the spawning place - indeed only for a short time. An actual territory is not estalished.

Since the revision by REGAN (1905), the gender of the generic name was considered feminine. But as KULLANDER states in his work from 1986, it is neutral so that following the rules in adjectival forms of the specific name (here in both!) the ending has to be changed from "-a" to "-um".

The generic name created by STEINDACHNER (also the following for *Dicrossus*) refers to the indent of the gill cover: ›*crena*‹ (lat.) = notch.

Crenicara latruncularium

KULLANDER & STAECK; 1990

Meaning of the specific name (Etymology): The name of this type derived from the trivial name that is not solely for the fish of this genus, but also for those deriving of the *Dicrossus*: ›Chess Board Cichlid‹. The Latin word *tabula latruncularia* is the translation for chess board, and the body pattens of the fish correspond with the checked pattern, especially when in an agitated state.

Remarks about generic gender: Cp. comments at the end of the genus review.

First description: Cybium 14 (2), pp 161-173.

Synonyms: None.

Distribution: As far as one can establish, the fish comes from the border between Brazil and Bolivia. This is the entry area of the two big rivers, the Rio Guaporé and the Rio Mamoré. One can catch the fish even in the upper reaches of the Rio Guaporé, and I already had a few fish in my net that looked like catfish as I was near to Pontes e Lecerda and Vila Bela (cp. MAYLAND, ›Diskusfieber‹ 1988, pg 184), but the Cichlids, who were then only known as *C. punctulatum*, received little attention.

At the place of collection, near Vila Bela there was luxurious plant growth, among them, the different *Echinodorus*- and *Cabomba*, *Pontederia*, *Salvinia* and *Nymphaea* species, as well as *Eichhornia azurea* and *Trapa brasiliensis*.

Habitat: Due to the extreme softness, and in many zones slow flowing or almost stagnant water (Total hardness 0,02!) the habitat is not only for catfish and small Cichlids (among them

S22815-3 *Crenicara latruncularium*, wild male.

also, *A. inconspicua* and *A. trifasciata*), there are also large Cichlid types for example, *Satanoperca pappaterra*, *Aequidens plagiozonatus* and one can play with the beloved river dolphins in a few places.

SD and colouration: To differentiate between this type from the further north found *C. punctulatum*, the authors gave the following characteristics: Number of dorsal spikes are (only) 15 and in *C. punctulatum* it is 16 to 17. A further characteristic that is especially marked in adult males is that one sees on the rust brown base of the cheek three to four sloping green to light blue iridescent running bands. Specimens of *C. punctulatum* do not have these bands! Females develop during breeding time, red ventral and anal fins and the before mostly as a blotched row appearing lateral band pattern turns into a wide black band.

Total length: Male 12 cm, female 11 cm.

Maintenance in the aquarium and reproduction: Similar to as described for *C. punctulatum*.

Biotope: Upper Rio Guaporé
near Pontes e Lacerda, habitat of *Crenicara latruncularium*
Date: 20.10.1986
Time: 16.30 o'clock
Water movement near the shore: sluggishly flowing
Water depth: Knee deep up to 1.20 m
Water colour: Rather clear.

Field measurement:

pH-value	6.9_{20}

Laboratory results:

Electrical conductivity (20°C), µS/cm	18

Biotope: Upper Rio Guaporé
In lagoons rich in vegetation on farm land approximately 15 km downstream of Vila Bela
Date: 31.10.1985
Time: Not noted
Water movement: Still moving river water
Water depth: Several meters
Water colour: Clear, but visibility was impaired by vegetation

Field measurement:

pH-value	7.3_{20}

Laboratory results:

Electrical conductivity (30.5°C), µS/cm	50
Iron (Fe), mg/l (Dupla quick test)	0.1-0.2
Phosphate (PO_4^{3-}), mg/l (Dupla quick test)	<0.1
Oxygen (O_2)	7.27 (fol. Winkler)
Water temperature	30.5°C
Air temperature	32.0°C

Crenicara punctulatum GUENTHER, 1863

Meaning of the specific name (Etymology): From the Latin word ›*punctulatus*‹ = spotted.

Remarks about the genus race: cp. comments at end of the genus review.

First description: Ann. Mag. Nat. Hist. (3) 12, p. 441.

Synonyms: *Acara punctulata* GÜNTHER, 1863; *Crenicara elegans* STEINDACHNER, 1875; *Aequidens madeirae* FOWLER, 1913; *A. hercules* ALLEN in EIGENMANN & ALLEN, 1942.

Distribution: The specimens that GÜNTHER used for the type materials came from the entry of the Essequibo River into Guyana (Gluck Island). Later discoveries recorded a find in the lower Rio Içá, alongside the middle and upper reaches of the Amazon, in the drainage of the Rio Maderia (with Rio Madre de Dios/Perú and Rio Beni/Bolivia), further in Peru in the drainage of the Rio Tambopata, alongside the Rio Ucayali (Pebas, lower Rio Nanai, Jenaro Herrera, Tournavista), In the Rio Marañon (Rio Morona) as in the Rio Javari (Colonia Angamos and near to the river mouth).

Habitat: One seldom sees the fish in large numbers. They live preferably in shaded forest streams, with a sandy ground, in black or mixed water. Because of the widely spread distribution area it is not possible for all of the biotopes to have the same characteristics. Therefore, the fish are not a problem to care for when they are in a middle to large aquarium.

SD and colouration: This larger species, in relation to the Dwarf Cichlids, shows a different gender development in its body form (fins, head) and colouration (sexual dimorphism). The ventral fins, as well as the anal fins are iridescent light blue with white marks in the anal area. Dorsal and anal fins are are extracted in the soft ray area. In the smaller female fish, it is predominantly the pectorals, however also the anal fins, that show a powerful striking red orange colouration. One can describe both the head and body base colouration of both males and females as a dark sandy colour. With females, the dorsal side is often darker and appears an olive colour. With males, the throat, chest, and ventral fins are gold, against which females at best show a greenish iridescent zone on the gill covers. The mood related "chess board pattern" already talked about, is shown by both sexes, only in the form of a right angled spotted row unbroken lateral band. If other fish show fear, a similar (camouflage) pattern can be seen over the whole body.

Total length: Male 12 cm, female 7 cm.

Maintenance in the aquarium and reproduction: Chess Board Cichlids love a habitat filled with plants, soft, lightly acidic and of course nitrate poor water, as well as a diet that is in no way limited to dried food. In roomy aquaria, one can add many females when there is a dominant male. The females who are not submissive will in turn be fought, without sustaining injuries that are too severe. The Cichlids remain normally shy when one does not add several other fish - especially no fish living in schools.

Female fish have normally a horizontal band pattern. This pattern changes to broken up forms of blotches as a sign of submission to the dominant partner. When there are many other fish in the aquarium, there can be many fights between them, leading to impressive broadsides, attacks on the caudals, and front on confrontation. The known dragging of the mouth typical of most all other Cichlid types has not been noticed, due to the small mouth of these fish.

When the female fish are ready to spawn, they do not make themselves more attractive, which is the normal behaviour for the fish of a different genus, but rather they hold back, and develop an insignificant grey colouring. The male soon recognises the situation and begins to show interest in the female. This interest, instead of beginning with tenderness, begins rather with an impressive show. The golden colouring in the throat, chest, and pectoral area intensifies. The time is almost ripe for the pair to search for a breeding place, in which the female takes the dominant position. A leaf or a stone can be chosen as substratum. To show that the breeding place is suitable for the female, she begins to clean the area with pecking movements. The male does not take part. After the synchronization of the reproduction ritual is complete, the female begins to lay her eggs. Depending on the size and bodily condition of the female,

S22820-3 *Crenicara punctulatum*, wild male.

it can take around 30 minutes to lay around 80 to 140 eggs. These are immediately fertilized by the male. This is the last interference that the female permits. She attacks her partner, and chases him out of the breeding area. She alone takes responsibility for the brood. The scorned male has no other choice than to take a new partner from a possible ›harem‹ and to begin the reproduction cycle again.

The larvae, with the help of the mother, free themselves from the egg capsule after 60 to 70 hours, and develop further, so that they can swim freely after seven to eight days. Unfortunately it is not possible to rear the full number of offspring (measuring from the amount of eggs laid). The red fins in the lower body region act as a signal to follow the mother in the search for food. Once the young have reached a length of about 2 cm, most of them develop red pectoral, and anal fins, a sign that we are exclusively dealing with females. Merely a few young grow faster and outgrow the rest of the group, showing that they are males. Their pectoral and anal fins develop a blue colouration.

Remarks about the sex transformation: Prof Dr Dietrich OHM from the technical University of Berlin, reports in different articles in the year 1980 about the different tests that he has carried out. »From the beginning is was noticeable that most of the fish would be female, and that only a very small number of males existed. In order to solve this problem, I placed 20 to 30 day old young in approximately 40 groups of about four fish together. These individually known groups were observed for over one year.«

Very seldom one can see old or young fish, in normal circumstances, with high positioned fins. Their fins mostly lie flat, which understandably, many aquarium hobbists do not accept, because the imposing impression of the fishes body is then lost. Regardless of the small mouth of this fish, and regardless of the described closed position of the fins, aggressive behaviour toward each other is not to be dismissed. For the superior fish in the aquarium, it is of highest importance to demonstrate their rank in the hierarchy, which is shown predominantly during feeding. The so called aggression lasts only a few seconds, but produces the desired effect for the dominant fish: The attacked fellow occupant pulls back from the immediate vicinity of the higher ranking fish. The establishing of territory, as known from other Cichlids throughout or during reproduction, is not seen in the *Crenicara* representatives. The hierarchy must be firstly ›defined in fights‹ with the young fish, or in a newly established group. These fights between the young begin with the age of six to eight weeks. Even with these young fish, it has shown that the superior fish must display a continuous lateral band, against which, the inferior fish show a different checkered pattern depending on the mood.

After the young fish have reached the age of about three months, the pectoral fins and the anal fins begin to develop a pink, and later a reddish orange colour. This is however, the typical colouring for females! Because the young fish do not all reach the same size, (a few specimens can grow remarkably bigger), but regardless the ranking order has already been established, the larger adults, and the higher ranking fish begin to show their male sex. These fish lose the reddish orange colouring on the pectorals and anal fins, and acquire a slightly iridescent, sky blue colouring. The fin tips extend increasingly, and the extended first pectoral fin ray acquires a pure white colouring. The posterior (soft) rays of the dorsal fins also extend and develop - as do the caudal fins - a spotted pattern on a pale blue base.

The mentioned author (Prof Dr OHM) reported that he could prove this phenonomon in more than fourty groups of fish, each group consisting of four fish. The specimens that did not transform into males, retained their female gender. The transformed males, who had originally shown female charateristics, began to reproduce with the left over females at a mature age of 10 to 12 months. Therefore, a sex transformation like the one described is based on social superiority. This is however not the exclusive trigger.

At the age of about two years it was still observed that in fifteen groups, the transformation from female into male was established. These females had lived before in the ›harem‹ of a male, and were fully functional as females and had also bred young. »Generally« - remarks the author - »It appears that the first females to transform are those who had spawned less. In many groups all that was left of the former harem, were males. Therefore the problem of too many males comes to the fore, and it is unsure if this same phenonemon occurs in their natural environment.

During their excursions from female to female, the relatively larger males can very easily fall victim to an enemy, or easily be driven downstream by the flowing water. It is then important for the preservation of this species, that one of the females transforms itself into a male, to adopt or take over the specific functions.«

The genus *Dicrossus*

STEINDACHNER, 1875

For many aquarists, *Dicrossus* is a new genus name, that however - one can see by the date - has existed for over one hundred years. It slumbered decades however, under the synonym of *Crenicara* (through REGAN 1905) and was ›revalidated‹ in the year 1990 by KULLANDER (In

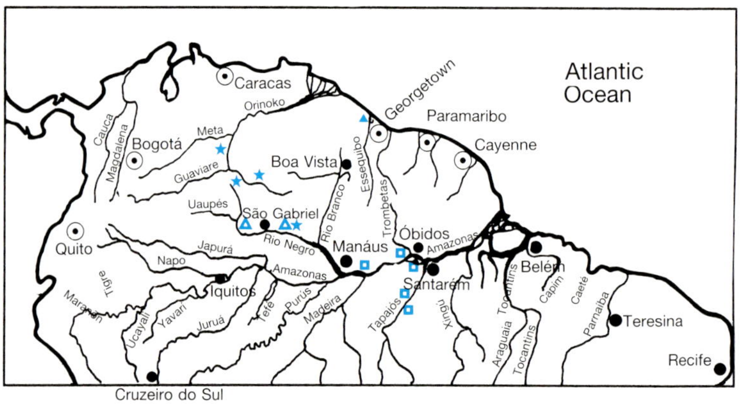

★ *Dicrossus filamentosus*
□ *D. maculatus*
△ *D. spec. „Zweipunkt/ Rio Negro“*
▲ Crenicara punctulatum (distribution outside of the Amazon basin; cp. also map page 159)

S31160-3 *Dicrossus filamentosus*, wild male.

the scope of establishing of the genus *Mazarunia*). Until then, in the genus, only two types were described, today we know of a further two species awaiting their scientific description. For many aquarists the combining of the genera *Crenicara* and *Dicrossus* from early on was not so clear, as the members of *Crenicara* have a much higher back, and the males who reach a length up to 12 cm are much larger. The chess board body pattern as well as certain behaviour patterns, in jerky swimming behaviour, and in reproduction rituals, are totally similar in both genera. In contrast to their *Crenicara* relatives, the members of this genus are less shy in well kept aquaria, but however as remarked, reproduction is not to be seen as ›something for beginners‹. One wanting to reproduce these small Cichlids, needs not only a lot of empathy, but also a very soft and very acidic water. The generic name is made up of the greek prefix ›*di*‹ = two, doubled, and also the greek word ›*krossos*‹ = fringe, tassle. This can only relate to the notches on the gill covers, which both the representatives of *Crenicara* and *Dicrossus* have.

Dicrossus filamentosus LADIGES, 1958

Meaning of the specific name (Etymology): From the Latin adjective ›*filamentosus*‹ = furnished with filaments. The author refers to the long extended, thread like membranes (filaments) on the upper and lower caudal fins.

Remarks: According to the international ruling on the adjective form of a specific name, the ending has to be adapted to the gender of the generic name, and because *Dicrossus* is masculine, the ending must read "-us".

First description: The article published in DATZ 11 (1) 1958, page 203-204 has priority: ›Remarks to several new imports‹. The actual scientific work was published one year later: International Rev. Ges. Hydrobiol. 44, pages 266-302 with the title: ›*Crenicara filamentosa*‹ spec.

nov., a new rare Cichlid from ›South America‹. In the year 1978, KULLANDER published a new description.

Synonyms: None.

Distribution: Because the aquarium breeding in this style is not easy, the fish are more commonly imported out of central Amazonia (Manáus). Many young fish, a few of these are still undescribed *Apistogramma* species, can be found in these imports. The fish of this type are widely distributed and probably live together with the Neon Tetra (*Paracheiroden axelrodi*). This distribution reaches from the entry of the upper Rio Negro near Barcelos, to the area of the Rio Vichada (a tributary, to the left of Colombia, of the upper Rio Orinoco) I myself (MAYLAND) did come across the fish in the vicinity of São Gabriel (Rio Negro) as well as in the area of the Rio Guaviare and the Rio Inirida (Colombia), and tributaries of the Rio Orinoco (cp. map page 164).

Habitat: In all the named areas, the water is extremely soft (cp. Water test ›Upper Rio Negro‹ in *A. diplotaenia*) and has an often above average acidity (between 4.5 and 5.5 pH). There is enough forest, surrounding the shallow and tea coloured water edges so one can find in the warm streams a lot of foliage and branches. Submerged vegetation exists very seldom, if at all.

There are two different patterns of caudal fins known in males. The pattern of the upper fin is shown by the specimens from the Orinoco region, whereas the one of the lower specimen occurs in the Rio Negro area.

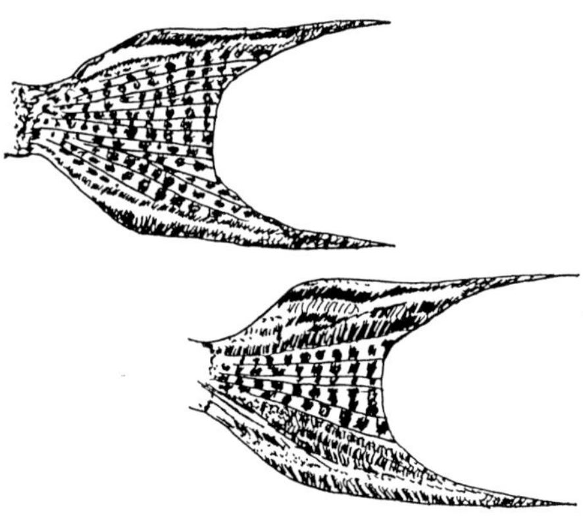

SD and Colouration: Because of the wide distribution, one should not be astonished at variations in the colour patterns. In fish of this species, it is especially conspicuous in the area of the male caudal fins: The fish imported from the Rio Negro and past Manáus have a relatively small wedge with a spotty red pattern in the centre of the caudal, marked off with a yellow rim inbetween. The species coming from the entry of the Rio Orinoco (introduced out of the Colombian capital Bogóta), has a far wider spotty wedge, and the yellow zone is hardly recognizable.

Virgin females have light coloured pectoral fins. When they reach sexual maturity, the colour changes to a soft red. After the first spawning the colour changes to a permanent colour of cherry red. One can then easily see if a female has spawned or not.

Total length: Male 9 cm, female 6 cm.

Maintenance in the aquarium and reproduction: Anyone who has thought over the given information about the water quality in the habitat of the slim Cichlids, should not be suprised to discover that the keeping, or rather the reproduction of this species is not an easy practice.

For the care and reproduction of this fish, one should provide a tank that is not too small. Tanks with an 80 and 120 cm front length, have proven successful. One (or for the time being) more males can be paired with several females. To reduce the shyness of the Cichlids in the tanks, and for better observation, it is recommended that one adds some accompanying fish.

The bottom is most functional when it is covered with a mixture of sand and fine gravel. In the habitats of South America we have seldom come across gravel, as known and used in aquarium circles. Sand with differing grannulation, mostly covered the ground (other than risen rocks, matured sediment and muddy particles). Because (almost) every keeper strives towards attempting reproduction, it is strongly recommended that one takes care both of the vegetation and the water in the aquarium. On most occasions, the females choose a horizontal lying leaf to lay their eggs. The introduction of wood or a peat filter can have a positive effect on the aquarium

S31160-3 *Dicrossus filamentosus*, female above the eggs.

environment. Conductivity should not exceed, if possible, 60 µS. Tests by experienced breeders have shown that the desire for reproduction is also influenced by the pH value in the water. It is often not enough to push the value under 6.0, but rather to achieve a value between 5.0 to 5.5. Also, a water temperature of 27 to 29°C is necessary. As one knows, it is not easy to hold these values constant for a long time, but because the stability of the water suffers, its value must be constantly controlled, and in emergencies corrected. The breeder should not acknowledge his success on the pink colouration of the female ventral fins alone. When both partners enter sexual maturity, they both show off their most beautiful colours. When there are many males in the tank, one recognises the dominant fish with a continuous wide black lateral band, that is edged from both sides by a light spotted row. In contrast, inferior males, instead of displaying a continuous band, show a blotched band.

When the time for reproduction has come, the female begins the related preperations with a massive cleansing of the selected substratum. The accompanying ›picking‹ can also be considered as an element of the mating ritual. The actual spawning and act of fertiliziation does not take longer than approximately 30 minutes. When the lay is complete, and the presence of the male is no longer necessary, the female attacks him in different ways to show him that his presence in the vicinity of her fry is not desired. It can also occur that the female actually eats her fry, and one must wait until the next spawning.

The larvae hatch after two to three days, and are transferred by the attentive mother several times a day. After a further three to four days, the young fish swim free.

Dicrossus maculatus STEINDACHNER, 1875

Meaning of the specific name (Etymology): Derived from the Latin word ›*maculatus*‹ and is the masculine adjective form. It means something like spotted.

First description: Sitzber. Kais. Akad. Wiss. Mathem.-Naturwiss. Wien 71 (1), pages 102-106.

Synonyms: *Crenicara m.*; *Crenicara praetoriusi* AHL, 1936.

Distribution: The type location was given by STEINDACHNER: Lago Máximo, Lago José Assú, tributaries of the Rio Tocantins, Rio Javari and Rio Tajapurú. On the other side, AHL names Igarapé Irurá-Mapiry. That means (and KULLANDER reached the same opinion in 1978) that the distribution includes the whole Amazon from Belém up to the Colombian boarder (Rio Javarí) With this in mind, it is difficult to understand why the fish has been so rarely imported. The modest number of specimens kept today based mostly on the merits of the Munich importer Arthur WERNER, the importer and breeder, Erich ›Peter‹ FRECH from Memmingen and the breeder Peter GÜNNEL from Zwickau (no others are known to us at present), who have gone to the trouble to catch the beautiful Cichlids south of the city of Santarém (WERNER) and to reproduce them (cp. map page 164).

Habitat: If the wide distribution of the species is true, then the biotopes would probably resemble *Apistogramma agassizii*. The waters

S31186-4 *Dicrossus maculatus*, wild pair.

alongside the Amazon equally reach the same low total and carbonate hardness all over (<1), as can be proved in the black water areas for *D. filamentosus*, and still the acidity of the water in the boarder areas of the Amazon is continuously constantly stronger. In areas of clear water rivers in the lower Amazon (Rio Xingú, Rio Tapajós and Rio Tocantins are the largest) the pH value can be very low. For the fish, this means that they have a large adaptation capability, and therefore the keeping and reproduction can be considered rather problem-free.

SD and colouration: The fish live up to their scientifically given name remarkably, because they show almost permanently two horizontal bands made up of large black blotches, one covering the middle centre, and a further underneath the base of the dorsal fins. In the standard colouring, males show iridecent blueish unpaired and ventral fins. The fins of the females are colourless. When the males are ripe for reproduction, the blue is far more intense, and predominantly in the dorsal and caudal fins developes a bright red colour. In the females, as in *D. filamentosus* that acquire red pectorals after their first spawning, one can see that the pectorals of *D. maculatus* are yellow.

In 1992, KILLIAN reported about the outdoor observations of *D. maculatus* that he made together with C. SEIDEL at the entry of the Rio Tapajós, in the further area of the city Santarém, near to the city Alter do Cháo (about 50 km away). The biotope connected with the large river in the fish rich Lagoa Jacundá had a white, sandy ground covered with rotting substratum a centimeter high and with crystal clear water (the Rio Tapajós is a so called clear water river). In different depths, the water had a temperature between 27 and 35°C. The given water value amounted to 10µS and 4,8pH. Total and carbonate hardness, as also the iron content, were not able to be measured. *D. maculatus* existed here »in giant numbers, even in real swarms. In groups from six to twelve fish, the Chess Board Cichlids moved over the ground and searched for something edible«.

For this, the fish stirred up the ground and picked at the edible particles, while lying flat on their stomachs on the ground.

Total length: Male 10 cm, female 6 cm.

Maintenance in the aquarium and reproduction: The keeping and reproduction is very similar to that of *D. filamentosus*. The males of *D. maculatus* are also polygamous, and can pair off simultaneously (but only in larger aquaria), with many females. It is then possible that in one aquarium, many females breed simultaneously - and all from the same father. The given reproduction data, for *D. filamentosus*, is similar for this species. Tests have shown also in *D. maculatus* that a very soft and rather acidic water is more productive towards the desire for reproduction in the fish, than a middle soft, and neutral acidic content.

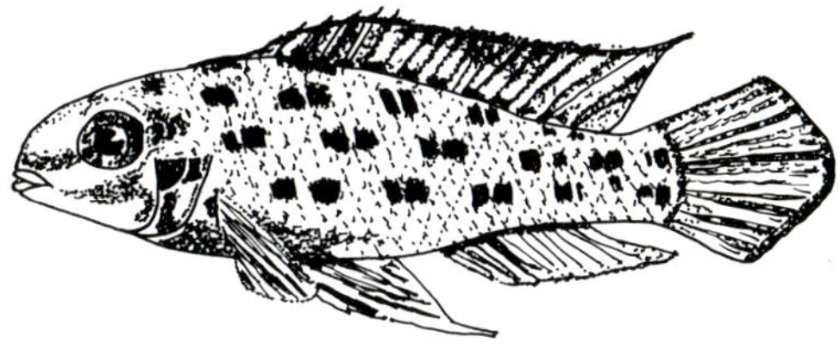

Dicrossus spec. ›Rio Negro‹, male.

Dicrossus spec. ›Doppelpunkt-‹ or ›Rio-Negro‹

New scientific name:

Author and year:

Distribution: In different areas with inflowing rivers into the middle Rio Negro. Approximately at the latitude of 61,4°W the Rio Padauari flows into the Rio Preto some 12 to 13 km upstream. The area to search in is after this mouth area, for it is here that the fishermen from Amazonia import the specimens to Manáus, the ›central sending station‹ (cp. map page 164).

Habitat: The tea coloured water is - as all around - very soft. The mineral content was not higher than 10 on the µS instrument. The pH value was in the very acidic area (5.5 to 4.4).

The habitats themselves are poor in vegetation regarding the above named conditions, and it is most of all the collection of wood (dead roots, and branches, and with this, foliage in abundance), that offers shelter for the fish. In their natural habitat, the fish live together peacefully with *A. regani*, *A. hippolytae*, *A. paucisquamis*, *A. gibbiceps* and *Dicrossus filamentosus*. Representing the *Crenicichla* family, *C. notophthalmus* and *C. regani* are present along with larger growing not yet identified species.

S31196-3 *Dicrossus* spec. ›Rio Negro‹, sexually mature wild female.

SD and colouration: Covering the females basic form, one can see a pattern of three horizontal rows of orderly black double blotches that lie behind each other. They disband over the same coloured head to a lesser marked blotch. With the male fish, the dorsal, caudal and anal fins are interspersed or rimmed with an irregular arranged red zone. The fins of the females are transparent.

Total length: Male 8 cm, female 6 cm.

Maintenance in the aquarium and reproduction: As to the aquarium, the fish prefer similar conditions as they know from their habitats. Therefore, they have no interest in vegetation, and like dim light. A lot of wood and a few built in caves for shelter is pleasing to them. The best choice for this are coconut halves and empty flower pots. Even in aquaria the size mentioned, the males set up and defend their territory, and it does happen - one can hardly believe it - that regardless of this, there are fights resulting in deaths.

For reproduction, this water should have a strong acidic pH value around 5.5 (as is normal for the typical genus relatives). In cases where the female has hidden her spawning, one sees her ›loss of virginity‹ in a shining red colouration of her pectorals. Alone the fact that *Dicrossus* types only spawn in very acidic water, is the decisive factor why one can acquire the fish only as wild fish and not as pedigrees. The female lays the eggs on a substratum (stone, leaf, depression in a root) and these are cared for by the mother alone. The pectorals of the caring parent fish do not colour up red after the first spawning (as with *Dicrossus filamentosus*) but rather yellow: The anal fins show red tones. At temperatures around 27°C the larvae hatch after three to four days. They are then transferred by the mother many times a day, and swim free after four to five more days. The young consume immediately small *Crustaceae*, for example *Artemia* larvae.

Dicrossus **spec.** ›Dreipunkt-‹ or ›Rio Tapajós‹

New scientific name:

Author and year:

Distribution: In different areas in the system of the Rio Tapajós (cp. map page 174).

Habitat: The Rio Tapajós belongs to the three largest clear water tributaries of the Amazon coming from the south(Rio Xingú, Rio Tapajós and Rio Tocantins). Warzel (Mayland, pers. comm.) reported about the appearance of Dwarf Cichlids in the Várzeas (= flood plains in the Amazonian river valleys) near to the Rio Tapajós. Half grown specimens are present here in large numbers, in the flat banks above a layer of foliage. Adult specimens are seldom seen here.

SD and colouration: As with the former species, these fish also have an almost pure white head and body, only belonging to the horizontally arranged double blotch pattern of the Rio-Negro-Dicrossus with the species coming from the Rio Tapajós because of three running flat black blotches. The males of this species also show red tones in the dorsal, caudal, and anal fins, so that the introduced popular name ›Rot-flossen-Dicrossus‹ (red fin Dicrossus) alone is therefore misleading (*D. maculatus* and *D. fila-mentosus* also show red tones in the mentioned fins).

Total length: Similar to the former species.

Biotope: Rio Tapajós,
above the city of Santarém, Pará (Brazil).
Date: 21.11.1996
Time: 9.15 o'clock
Water movement: In the wide delta area (into the Amazon) lake like width, and accordingly stagnant
Water colour: Clear, greenish if seen from above
Surface temperature: 26.8°C

Laboratory results:

pH-value	6.49
Electrical conductivity(25°C), µS/cm	22
Acid capicity up to pH 8.2 (p-value), mmol/l	n.p.
Acid capicity up to pH 4.3 (m-value), mmol/l	0.17
As carbonate hardness, °d	0.48
Total Alkalis, mmol/l	0.063
As total hardness, °d	0.36
Sodium (Na^+), mg/l	1.0
Potassium (K^+), mg/l	1.0
Calcium (Ca^{2+}), mg/l	1.60
Magnesium (Mg^{2+}), mg/l	0.57
Total Iron (Fe^{2+}/Fe^{3+}), mg/l	0.047
Manganese (Mn^{2+}), mg/l	<0.002
Ammonium (NH_4^+), mg/l	0.05
Nitrite (NO_2^-), mg/l	<0.02
Nitrate (NO_3^-), mg/l	0.6
Chloride (Cl^-), mg/l	0.8
Sulphate (SO_4^{2-}), mg/l	0.6
Phosphate (PO_4^{3-}), mg/l	<0.04
Hydrogen Carbonate (HCO_3^-), mg/l	10.4
Oxidation	
a) as $KMnO_4$ consume, mg/l	6.5
b) as Oxygen (O_2), mg/l	1.6

n.p. = not provable

S31198-3 *Dicrossus* spec. ›Rio Tapajós‹, young sexually mature male.

Maintenance in the aquarium: These fish also desire similar biotopes as they know from their habitat. The water of the Rio Tapajós comes from central Brazil. The river mouth is by the city of Santarém in Amazonia. On its long, middle and upper reaches, it takes its way mainly over the grannit or sand stone existing on the central Brazilian shield, that is often interspersed with attractive rapids. The bottom of this main river, as do its tributaries, consists of sand bars that in certain places mount up into sandbanks. In its lower regions, the river leaves its hard ground, and enters the Amazonian lowlands, and its bed widens to 15 km. Long islands build up on the river bed itself (similar to Anavilhanas-Island in the Rio Negro). The current developing in the delta bay sinks to approximately 1cm per second in the dry periods, so that conditions develop to that of a lake rather than a river. The then exposed shore of the bay of the Tapajós exists purely of light sandy beaches and in the water, one can see the build up of phytoplankton, the well known beginning of every food chain.

S31198-3 *Dicrossus* spec. ›Rio Tapajós‹, young female.

In 1983, the famous limnologist Dr. SIOLI published some tests of the water, where he was able to establish that both are extremely soft and that the water of the Amazon has a pH value (6.5 to 6.9) that is only slightly different from that of the Rio Tapajós (6.4 to 6.65), the existance of bicarbonate-CO_2 mg/l in the Amazon swings between 8.8 and 17.3, whereas the Rio Tapajós has shown only 3.1 to 5.2. That means, for the Rio Tapajós, that nutrients for plant growth are hardly present, neither in the shore area (pure sand floor) nor in the water (too little CO_2) (cp. actual water test).

The fish of this region also show no particular interest in plants, and therefore prefer a ›sunny‹ aquarium. A lot of wood and self made cave type shelters are enjoyed by these fish. The best material is again coconut shells.

Reproduction: Even though, according to WARZEL's statements, successful reproduction of this species in the aquarium has not occurred, there are impressions about the behavioural care of the fry that have stayed, regardless of the unfortunate half successful breeding attempts. As in the remaining *Dicrossus* relatives, the females spawn, if possible, on a hidden substratum - a horizontal leaf, root or in emergencies on a stone. Mating activities are mostly seen in the female. Her pectorals develop a red colouration and the two lower rows of blotches develop into an irregularly rimmed lateral band. In connection with the fertilization, the male shows no further interest in caring for the young, therefore leaving the female with sole responsibility. The female shows her revenge with increasing aggression. A reproduction attempt, at 26°C, a total hardness of about 1°d (conductivity 80 µS) and a pH value of 5.2, showed that after eight days, the young were so far developed that the mother could lead them around for their first feeding. Because in the first few days the mouth of the free swimming *Dicrossus* fry is not wide enough to eat *Artemia* nauplii and other small kinds of food in the amount needed was not available, the number of young dropped daily from an estimated number of 40 young, although in the old established aquarium, enough small organisms were present. When after 4 days the first *Artemia* larvae were consumed, these young fish had no further chance of survival.

The genus *Mazarunia* KULLANDER, 1990

An until now, in the aquarium, extensively unknown monotypic genus from the water section of Guyana, the former ›British Guyana‹ in the north-east of South America. The specimens that KULLANDER had at his disposal, for his genus and type description, came from a collection of Patrick DE RHAM in October 1976 from the upper Mazarun river. The author (KULLANDER) stressed that already in 1982 he realised the species belonged to a new genus, but a scientific description was not possible at the time because only two specimens were present. Because the new type is as important in many ways as the characterization of the genus *Crenicara*, the scientific description was carried out in 1990 regardless of the limited material. The description of a second type is being worked on (?).

What is Guyana? It is the westernmost of the three Guayana countries in north-east South America. In the north-west, the land borders to Venezuela, in the west and south to Brazil, in the east to Surinam and in the north to the Atlantic Ocean. One surmises that the name of the country comes from the Indian tribe of the Guainázes (›respectable people‹). Guyana is a flat and hilly land with a colonial heritage, which is considered a burden today.

At the beginning of the 17th century, Dutch trading posts were settled in the region of the Essequibo and later the Berbice Rivers, making the English and the French contentious towards the Netherlands, and in 1814 England had the upper hand and decided to add the two named areas along with the Demerara region, to colonies of British Guiana. British planters

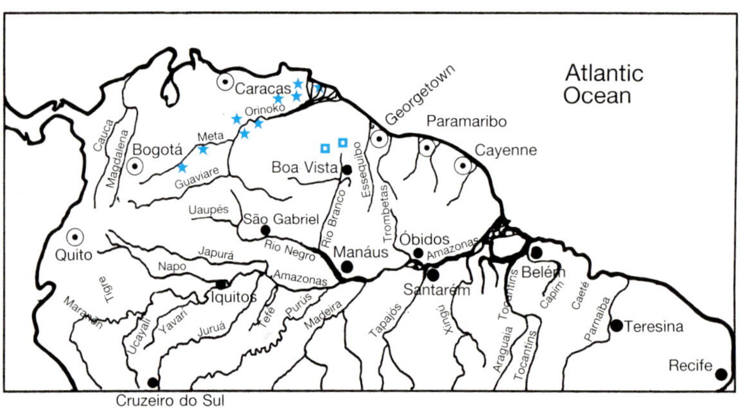

★ *Microgeophagus ramirezi*
□ *Mazarunia mazarunii*

S49700-4 *Mazarunia mazarunii*, young wild male.

imported, at this time, slaves from Africa to work for them on their sugar plantations. After the abolition of slavery (1838) one added immigrants from Portugal, Canton and China and hundreds of thousands of cheaply hired Indian workers to work in the British crown colony. In the country of this people and an explosive race mixture, in that next to the Indian, whites and Chinese minorities, 50 percent were Indian and 40 percent of African origin, it is difficult in European style, to maintain a certain order.

In the west and south-west, the Roraima Massif rises up out of the rainforest, which here still covers nine tenths of the ground area, with heights of up to 2772 meters. It is built up of hard sandstone, and quartz, that does not affect the water hardness. Numerous streams and rivers drop into waterfalls, including the 251 meter high Kaieteur-Falls. At the border of the massif, the wide plains, the so called Guyana shield, stretch out. The best time to travel is in the relatively dry months, September and October. Hotels are ›a scarce commodity‹.

S49700-4 *Mazarunia mazarunii*, young wild female.

Mazarunia mazarunii

KULLANDER, 1990

Meaning of the specific name (Etymology): Possessive form from Mazaruni = ›the Mazaruni‹ or ›coming out of the Mazaruni‹.

First description: Ichthyol. Explor. Freshwaters 1 (1), pp. 3-14.

Synonoms: None.

Distribution: At present, found only out of the Mazaruni river in Guyana. The specimens came from around the area of the small town Kamarang. The holotype is measured with a standard length (SL = measured not including the caudal fin!) of 53.4 mm, and the paratype had a given SL of 47.5 mm. One can then estimate a whole or total length (TL) of 8 cm.

Habitat: The procured specimens by the Dutchman Frans VERMEULEN came from the direct vicinity of the type locality, a sandy area of the Mazaruni river, three rapids down from the small town Kamarang, that lies in a higher triangular river mouth area. Indians make up the highest percentage of the population, rubber profits and panning for gold are the basis of their earnings. The Mazaruni river and its tributaries come from the near lying Merume mountains. Its soft water flows powerfully and has rapids in many places, many leading into waterfalls. The water ground exists of pure sand with occasional stone resp. rocky supports. Wood and vegetation is mostly missed here. The biotope of the small Cichlids, near a rapid with the name "Sand landing" exists of holes in the sand and caves in stones in the very shallow side areas of the river.

SD and colouration: Here we are dealing with not so colourful specimens, as in the related *Crenicara* species (photos). Males and females do not differentiate much from each other, merely the fins of the males become slightly more luxurious as the fish gets older. Females stay slightly smaller.

Total length: see ›Distribution‹.

Maintenance in the aquarium and reproduction: These shy fish with the small mouth and partly blue shimmering fins, as already mentioned, have certain simularities with the two *Crenicara* representatives. Their behaviour corresponds (as far as we know) with these relatives. Reproduction was not possible as of yet with the few imported specim

The genus *Microgeophagus* FREY, 1957

›Schmetterlingbuntbarsche‹ (Butterfly Cichlids), as they are known with their German name, have an interesting history as hinted at by their name, that is talked about at the end of the paragraph ›The Genus *Apistogramma* (A Historic Review)‹. They come from the area along the middle and lower Orinoco in Venezuela and Colombia (*M. ramirezi*), out of the eastern lowland of Boliva (*M. altispinosus*, ›one spot‹) and near to the boarder areas on the Rio Guaporé in the Brazilian Mata Grosso (*M. altispinosus*, ›two spot‹). In connection to this, it is interesting that at the beginning there was a lack of clear informative views from the former leader of the Fischbestimmungsstelle in the VDA, Hermann MEINKEN. Today many aquarists who keep the fish cannot understand his

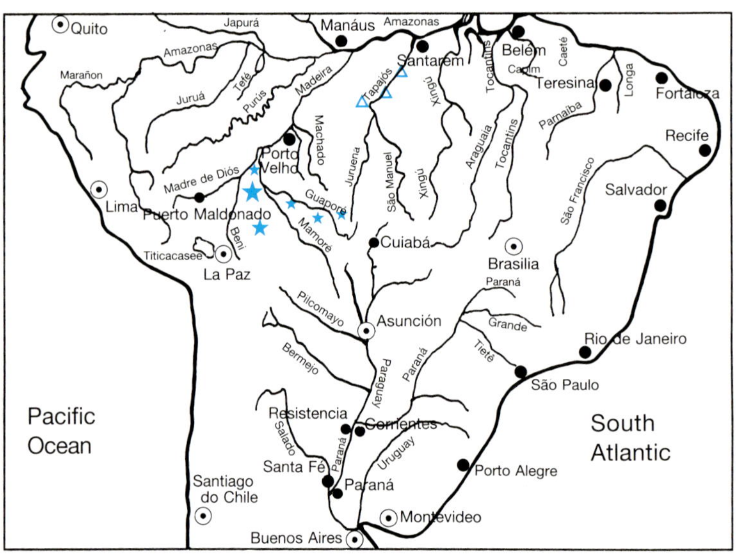

★ *Microgeophagus altispinosus*
△ *D. spec. „Dreipunkt/ Rio Tapajós"*

opinions or reproaches and therefore differentiate the representatives of both types purely on an optical basis.

MEINKEN regards the specimen supplied by Dr LÜLING, the ›Bolivian Butterfly Cichlid‹ incorrectly as a specimen of the type *A. ramirezi*, whereby he wrote: »...It concerns itself factually on a fully or almost fully grown male from *Apistogramma ramirezi* MYERS & HARRY, 1948, that differs only slightly in measurements from the original descriptrion. The differences are so small, that therefore no even the nomination of a subspecies or a set outlined variety were appropriate«. At the end of his publication, MEINKEN allowed himself the following statement: »It can be accepted as true, that Ramirez and Blass made a 500-mile-journey in Venezuela. This is not difficult to accept from them. But they did not catch *Apistogramma ramirezi* in Venezuela!«

A correction was later successful after harsh criticism. Today we know that there is considered a second type of the Butterfly Cichlid out of Boliva, that HASEMAN in 1911 placed in the genus *Crenicara* (as the present generic variety did not exist).

The three genera, which till now the small Cichlids have fallen through, caused many ›simple thinkers‹ to have headaches. *M. ramirezi* was described as *Apistogramma* and transferred to the genus *Papiliochromis* by KULLANDER that he created in 1977. It must also be said that earlier other generic names were published, whose validity struggled, because they were not based on or accompanied with sound scientific work (One asks if the International Nomenclature Rules have a ›hole‹?) Over the years, the also by ichthyologists (e.g. Gery.) supported view, that the proposed generic name *Microgeophagus* from Frey, 1959, had priority over *Papliochromis* was accepted. In 1986, the Ichthylogist ISBRÜCKER from the Netherlands wrote: »according to the nomenclature rules (Paragraph 11, 13 and 15) this method was a perfectly proper way to set up a new genus, and in fact, according to the rules up until 31. December 1960. The main question is whether *Microgeophagus* FREY, 1957, must be considered as nomen nudum (naked, invalid name) or not.«

In an article about this theme, van den NIEUWENHUIZEN (1990) points out the following facts: »*Microgeophagus* was set up as a subgenus (of *Apistogramma*), and in fact with an explicit given type name. Paragraph 15 of the code states that after 1960 this procedure is no longer admissible. Therefore it follows that the name *Microgeophagus* is correctly set up and valid!« Whether this opinion is undivided, will be established with time.

The generic name comes from the Greek words, ›*mikros*‹ = small, ›*gea*‹ = earth, and ›*phagos*‹ = devourer (= small earth eater). The gender is masculine, whereas the adjective suffix of the earlier ›*altispinosa*‹ must actually read "*-us*". The genus belongs to the related circle of the *Geophaginii*.

Microgeophagus altispinosus HASEMAN, 1911

Meaning of the specific name (Etymology): From the Latin words ›*altus*‹ = high and ›*spinosus*‹ = prickly. The name refers to the high spikes on the dorsal fins.

First description: Ann. Carnegie Mus. 7 (3-4), pp. 344-345.

Synonyms: *Crenicara altispinosa*; *Papiliochromis altispinosa*.

Distribution: The biotopes in which the ›Bolivian Butterfly Cichlids‹ had developed, lie far away from *M. ramirezi* from Venezuela and Colombia. The earlier mentioned material collected by Dr LÜLING (a single specimen), came out of east Bolivia, more clearly said, out of the surrounding land around the small town of Todos Santos, at the edge of the low lands, whose waters belong to the drainage of the Rio Ma-

S52015-4 *Microgeophagus altispinosus*, ›Two spots‹, juvenile wild male.

moré. The region still belongs to Amazonia, because its waters flow into the Amazon through the Rio Guaporé and Rio Maderia. Further finds were made out of a lake west of San Joaquin, in Rio Quizer (14 km from San Ramón in the direction of Limón) as in the area of the boarder area of the Brazilian lying city, Guajará-Mirim. These exports refer to the statements made by former King Léopold III. from Belgium, and Belgian ichthyologist Dr J.-P. GOSSE, when they journeyed here. In new searches, the authors LINKE & STAECK (cp. 1986 resp. 1987) "struck gold" in the area of the city Trinidad in North Bolivia. The increase of the agricultural areas and with that, the excessive use of chemicals, destroy many habitats of the small Cichlids.

S52005-3 *Microgeophagus altispinosus*, ›One spot‹, young wild female above the eggs.

S52005-3 *Microgeophagus altispinosus*, ›One spot‹, wild male.

Remarks: A form with a varying pattern was imported for the first time at the end of 1992 out of the Brazilian Mato Grosso, near to the Bolivian boarder. Those fish (photo) with an impossible to miss large black blotch over the whole caudal peduncle, came from the Rio Pindaituba, a tributary of the Rio Sararé in the network of the upper Rio Guaporé, east of the city Vila Bela da Santissima Trinidade, and north of the township Pontes e Lacerda (cp. MAYLAND, 1995[b], water tests cp. page 174).

Habitat: In more or less shallow water types (mostly pools). L. & S. give a water temperature of 27.1°C, a total and carbonate hardness at 4°d, a electrical conductivity of 123 µS and a pH value of 7.6 for a similar biotope.

SD and colouration: The larger males are more intensively coloured, and show a luxurious finnage when they are almost adult. The male sex can be easily recognised by the difference in the caudual fins which shows a lengthening in the upper and lower red ray.

Total length: Fully grown males 8 cm, females slightly smaller.

Maintenance in the aquarium and reproduction: One places many specimens in a aquarium of 80 to 100 cm long, and leaves the pairing up of the fish to themselves. A ground out of fine gravel (fine sand is better, as the most biotopes in Amazonia have a sandy ground) and roots and stone shelters can be added along with vegetation, for the benefit of the fish. The water value should orientate itself on that of the already described.

As coinhabitants, catfish (*Corydoras*) are recommended. A partial water change of about 30 percent every two weeks allows the water to retain its optimal quality.

The pairing up of the fish, is recognised in the following behaviour patterns: After a period of apparent tolerance towards the other specimens, the behaviour, especially of the strongest male that has already chosen a female, turns rather rough. The attentive observer can soon see the staying together of the two. The establishment of a territory occurs as the pair receed to a chosen area in the aquarium and defend this area aggressively. Often an area is chosen which was prepared by the keeper, a flat pebble or a dark stone plate that has a size of about 10 x 10 cm.

Mating occurs after a few days, or sometimes after a week. During this time the male constructs, as synchronisation of the later spawning and fertilisation ›games‹, around the substratum, small approx. 2 cm deep hollows. When the female feels the beginning of the spawning time, she begins to actively clean the substratum of the chosen, approximately 4 x 4 cm large area.

The male plays no part in this work, but rather chases other fish out of their territory. When the female has laid her eggs, the partner shows up, and fertilises the separate spawn push for push. It can happen that overzealous males can ›forget‹ to fertilise the spawn. When such a thing happens, the eggs will be gone the next day, and one must wait until the next lay. In such a situation, one must provide the female with a nourishing diet for the following days.

The care of the fry is taken over by the female, who places herself tightly over the eggs and fans oxygen rich water over them. It is also possible that the male cares for the eggs for a short period of time, but this does not last too long. The larvae free themselves after three to four days at a water temperature of 26 to 27°C. They are then led by both parents over the hollows prepared by the male, and often transferred. Here the male also takes a part in the care and this can also lead to quarrels beetween the parents.

It lasts 8 to 9 days before the young fish have passed the larvae stage, and can freely swim. In the beginning, the young live off their yolk sack, and soon the bigger young fish can consume *Artemia* larvae. Dried food of any form - including powder form - is not recommended! In leading the young, the parents are themselves especially attentive, and chase vehemently all curious members away from the vicinity of their young.

Microgeophagus ramirezi

MYERS & HARRY, 1948

Meaning of the specific name: In dedication or in honor of the importer, Manuel V. Ramirez, who was the first man to introduce the fish in both the USA and to science.

First description: Proc. Calif. Zool. Club 1 (1), pp. 1-8.

Synonyms: *Apistogramma ramirezi*, *Papiliochromis ramirezi*.

Distribution: The origin of the first specimens of the >Butterfly Cichlids< was kept a secret - possibly to keep the commercial exploitation of the beautiful Dwarf Cichlids as confined as possible. The earliest information about the biotopes was found in a DATZ-report by Paul E. FISCHER (1968) out of the Venezuelan Caracas, in which the author and his guide were able to catch the first fish of this Dwarf Cichlid type, between the townships Calabozo and Paso de Caballo (Almost 120 km north of San Fernando de Apure) at the entry of the Rio Guarico and its tributary (Rio Orituco)

Today we know the widely stretched distribution along the lower and middle Rio Orinoco (in Venezuela) as in further regions along the Rio Meta (Colombia), also approximately between 4. and 9.°N as in the 62. and 72°W. They actually come up to the edge of the Orinoco Delta! Here we also met colourful populations, (from the north) actually half way between the towns Maturin and Tucupitá (cp. Map page 172).

Habitat: *M. ramirezi* is no (!) inhabitant of idyllic rainforest streams, rather it inhabits much preferably the savanna pools in the Llanos of the given low lands, where the agriculture is based on cattle raising and breeding. The pools origi-

Biotope:
Venezuela, Rio Cunaviche.
The river traverses the Llanos from west to east and flows into the Rio Arauca south of San Fernando de Apure, which on the other hand flows into the central Rio Orinoco.
Date: 19.2.1991.
Time: 11.00 o'clock.
Water movement: sluggish.
Water depth: 1 to 2 meters.
Water colour: slightly brownish with a green shimmer (see above).

Laboratory results:

pH-Value	6.79	Magnesium (Mg^{2+}), mg/l	1.2
Electrical conductivity (25°C) ,µS/cm	40	Total Iron ($Fe^{2+/3+}$), mg/l	0.14
Acid capacity up to pH 8.2 (p-value), mmol/l	<0.05	Manganese /Mn^{2+}), mg/l	0.049
Acid capacity up to pH 4,3 (m-value), mmol/l	0.38	Ammonium (NH_4^+) mg/l	0.11
As Carbonate hardness, °d	1.1	Nitrite (NO_2^-), mg/l	<0.02
Total Alkalis (Ca + Mg), mmol/l	0.13	Nitrate (NO_3^-), mg/l	<0.1
As total hardness, °d	0.8	Chloride (Cl^-), mg/l	0.60
Sodium (Na^+), mg/l	1.8	Sulphate (SO_4^{2-}), mg/l	0.85
Potassium (K^+), mg/l	1.5	Total Phosphate (PO_4^{3-}), mg/l	0.04
Calcium (Ca^{2+}), mg/l	3.4	Hydrogen Carbonate (HCO_3^-), mg/l	23
		Oxidation	
		a) as Potassium Permanganate $KMnO_4$ consume, mg/l	21
		b) as Oxygen (O_2), mg/l	5.3

S52020-3 *Microgeophagus ramirezi*, adult male of the blue pedigree.

nate often from used man made watering holes, and are surrounded by palms. According to the fact that most of the waters that *M. ramirezi* are collected in were prospected during the dry season (spring) the water temperature in the biotopes was relatively high. On the average, it was at 30°C, but it can also reach higher temperatures. During the European summer, in northern South America, it is only a few degrees colder with a slightly higher water level. The total hardness is in every case extremely low (about 5 to 35 µS) and the pH value lies in the very acidic area between 6.3 and 5.5.

The mostly sandy ground can produce plant growth (*Nymphaea*, *Eichhornia*, *Cabomba furcata*, *Eichhornia azurea*, *E. heterosperma*, *Heteranthera*, *Marsilea*, *Mayaca*, *Sagittaria* and others), but this is not necessary. Overhanging shore vegetation and palm foliage that has fallen into the water, is in their natural habitat often the only shelter for the fish. As coinhabitants, one sees many *Micropoecilia picta*, *Apistogramma hoignei*, *Hemigrammus armstrongi* as in a different type of Characin that reminds one of the *H. caudovittatus* coming from further south.

SD and colouration: See photos. Females ready to mate develop a red stomach. Next to the most intensively coloured wild form (pedigrees show appearances of degeneration) there are different, mostly in South Asia developed pedigrees, most famous is the ›Gold coloured Ramirezi‹. Female specimens stay slightly smaller and the males' long stretched out dorsal fin rays are remarkably shorter.

Total length: Male 5 cm, female slightly smaller.

Maintenance in the aquarium and reproduction: For this very beautiful and therefore very popu-

S52021-4 *Microgeophagus ramirezi*, wild female Orinoco delta.

S52021-4 *Micro-geophagus ramirezi*, wild male Orinoco delta.

lar Dwarf Cichlid species, one needs a tank of 50 to 60 cm length for the keeping of pairs. Also when, as described, in the habitat of the fish, plants are not seen all over the place, it is recommended that for the care of these fish that one has dense vegetation as it gives the fish more sense of security. It is already known that Butterfly Cichlids do not belong to the fish who breed in caves, but they still need their retreats and their hiding places.

The bottom should consist of fine sand, with further introduction of roots. As a special tip, for both the fish and the observer, one should lay a flat and when possible, dark pebble of about 5 x 10 cm large in the foreground of the tank. The female will then most likely use this as spawning substratum.

Because the majority of specimens offered in the shops are offspring, and therefore already well adapted, the keeper (purely for the keeping of the fish!) does not have to keep so strictly to the given natural water values. Also the wild caught fish can be well kept in water considered ›of average hardness‹ and ›neutral‹. The same is true for the aquarium temperature; as for most species, also here the usual figure of 26°C is recommended.

In contrast to this, if the fish should reproduce one must understandably adjust the water conditions closer to those of the natural habitat, so that a normal development of the eggs and the larvae is guaranteed. Such an adaptability is known in the breeds of many other South American Cichlids (e.g. the Discus fish), because under different circumstances, alone the osmosis can destroy the eggs.

In contrast to most of the *Apistogramma* representatives, the fish of this genus establish a parent family. Both parents care equally for the young. If the female does not accept the provided pebble substrate, she finds a small hollow in the sandy ground to lay her eggs in. After hatching, the larve are often transferred. When the reproduction temperature reaches 29 to 30°C, the larvae will reach the end of their development stage after six to seven days, and swim free as young fish. They are led around alternately by both parents. Because the mouth of the fish at this time is not so developed that they can they can consume the normal *Artemia* nauplii, the breeder must provide a finer substitute diet at the right time.

The genus *Taeniacara* MYERS, 1935

ARNOLD reported already in 1914 about the new type that he introduced as ›*Apisto-gramma* spec.‹. Further specimens occurred in the USA in 1935. Unfortunately the earlier communication between the scien-tists was not as developed as today. So it happened, that after the creation of the genus and the discription of the species by MYERS in 1935, the Berlin

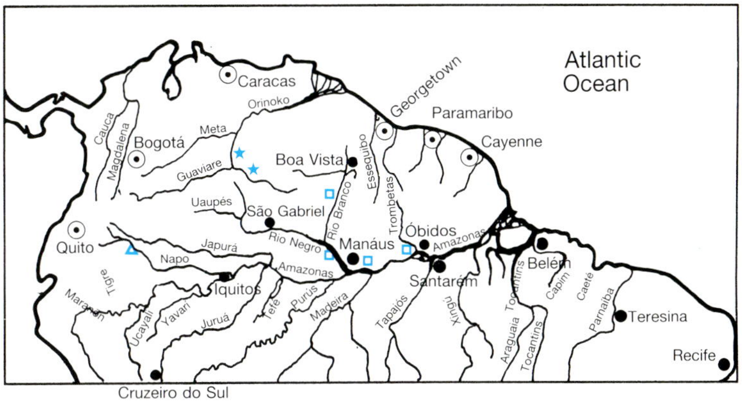

□ *Taeniacara candidi*
△ *A. payaminonis*
★ *A. spec. „Orangesaum"*

ichthylogist Dr. Ernst AHL in the year 1936, produced a subsequent second description. He created therefore, no new genus, but rather he added the new type to *Apistogramma*. ›Weise's Dwarf Cichlid‹ (*A. weisei*) was given a detailed description in the work of ARNOLD & AHL (1936, page 470-471). As we know today, this description is in fact a synonym.

The first introduction of the fish of this type was actually in 1913 successful (after A. & A.) The generic name comes from the latin words ›*taenia*‹ = band, and ›*acara*‹, a generic term for South American Dwarf Cichlid. It is supposed to refer to the impossible to miss, wide horizontal or lateral band.

The single representative of the genus *Taeniacara* differs from the relatives of the genus *Apistogramma* in the missing epibranchial lobes and a lateral line.

Taeniacara candidi MYERS, 1935

Meaning of the specific name (Etymology): A dedication name, or in honour of E. CANDIDUS, whom the author thanks for his type material.

First description: Proc. Biol. Soc. Washington., Vol. 48, pp. 11-13.

Synonyms: *Apistogramma weisei* AHL, 1936.

Distribution: The specimens were mostly introduced out of Manáus as fish caught with others. The small Cichlids coming from AHL's type material seem to be rather dispersed, from the

S92010-4 *Taeniacara candidi*, adult wild male.

central and lower Rio Negro, and the central Amazon (predominantly the left side) to the surrounding area of the city Santarém. In Linke & Staeck (1984/1995) the distribution area is enlarged and include the upper Rio Branco (?)(cp. map above).

Habitat: In small shallow biotopes rich in shelter near the banks. The water is extremely soft and totally nitrate free. Its pH value can change slightly, but stays very acidic. The ground covered with vegetation is interspersed with roots, and structured throughout with fallen leaves. The information about the Rio Negro should not bring the reader to the idea that *T. candidi* is solely an inhabitant of black water. One finds the fish also in clear water and - in the area of the city of Santarém - also in muddy habitats influenced by white water.

SD and colouration: Differentiating the sexes is very simple: The very slim male specimens have a large lancet-like pointed and long extended caudal fin, as with strongly extended tips on the ventral fins. In females, the caudal fins are rounded. In relation to their much more colourful partners, the colouration of the female is very much decreased. The lateral band from which the genus receives its name, exists in both sexes. It can be supressed in specific moods, but is mostly easily recognised.

Total length: Male 7 cm, females slightly smaller.

Maintenance in the aquarium: The specimens that are not so problem free are kept in pairs, or several females are added to one male. As cohabitants one can introduce small *Corydoras* or *Aspidoras* species, also recommended is a small group of slim Characins living in swarms (e.g. *Hemigrammus bleheri* or *Petitella georgiae*). The aquarium should have a front length of 70 to 80 cm and a sandy ground with a range of low growing plants (if applicable set in pots). Perfection is complete with added roots or stones for shelter. More than for other fish from Amazonia (e.g. the Neon Tetra *Paracheirodon axelrodi*), for the care of these fussy small Cichlids a soft to middle hard (most importantly carbonate poor water) and extremely acidic to neutral water is necessary. The adoption into a certain water type must happen with extreme care and when one changes the water, one must observe that the water quality remains the same. Through our experience we have seen that many specimens react sensitively to large amounts and to sudden water changes. The regular partial water change must be then performed slowly and with a certain amount of caution. The aquarium temperature should be about 26°C.

Reproduction: What was just said for the water value, must be considered in a much higher capacity in reproduction: The living element of the fish, and with that the development element of the eggs should be very soft and most importantly, must be very acidic. Total and carbonate hardness should not exceed 1 or 2°d, which is also particularly important for the carbonate hardness. The pH value must be at 5.0 and therefore kept so low that it is even more difficult to buffer it (peat filtering). When the pH value increases, the eggs can be damaged.

There are not many breeders having experienced successfully the problem free reproduction of *T. candidi*. Their experience is similar even if the infomation given is not necessarily identical or suitable. There are keepers, who have had reproduction success in moderately hard and neutral tap water. That sort of success one should not consider a good example.

The reproduction temperature should be between 26 and 28°C, which also near the lower limit corresponds with the normal temperature. In pure reproduction aquaria, it is better to have few cohabitants if any at all. Then the aggression of the males cannot come to the fore.

The typical body rocking, combined with slight lowering of the head, starts the ›question game‹ of the male to the readiness of the female to spawn. When the rocking is returned, then the two move soon to the female's cave and the coupling begins. The eggs are then attached to the roof of the cave, and are immediately fertilised by the male.

Then the caring for the fry is the female's duty, after three to four days she transfers the slippery larvae and hangs them on the side wall. After five to six more days, the mother appears in front of the cave and leads her children to their first ingestion. They consume *Artemia* nauplii immediately.

Since the late sixties, Hans Joachim Mayland is an acclaimed author of many specialist books on ornamental fishes. Having published more than fifty books at the Landbuch Verlag alone, he has sold over one million copies of his books in the German speaking area. On uncountable journeys to tropical areas all over the world, he has collected not only general impressions and interesting photographic material but also valuable water parametres from the habitats where the ornamental fishes come from we tend in our aquaria at home. Working together with his long-time companion, Dieter Bork, with whom he also compiled photos and text for the "Große Aquarienpraxis, Band I" (1976), Hans Joachim Mayland is now able to present this very special book, the result of a fruitful friendship and cooperation.

Dieter Bork, born in 1945 in Hanau, since the year 1956 is a commited aquarist. His aquaristic experience was interrupted through his service in the army as was his following studies as Dipl.-Ing (FH). He works as an engineer, and is responsible for the procurement and maintenance of the production plants.
Since 1972 he is a member of the German Killifish Association (DKG). For over 25 years he is intensely occupied with the reproduction of fish in the aquarium. His special interest within the Killifishes, lies with the Dwarf Cichlids, rare viviparous as in different Dwarf Rasbora species. Since 1988 he takes regular collection and study journeys in the lands of origin of the aquarium fish, resulting in further activities in the form of lectures and publications in specialised magazines.

BIBLIOGRAPHY

Agassiz, J.L.R. & Agassiz, E., 1868: A journey in Brazil. Ticknor & Fields, Boston, 540 pp.; reprinted by Praeger, New York, Washington, London 1969.

Ahl, E., 1931: Neue Süßwasserfische aus dem Stromgebiet des Amazonenstromes. Sitzber. Ges. natf. Freunde Berlin, pages 206-212.

-, 1936(a): Über eine kleine Sammlung von Süßwasserfischen aus dem Gebiet des Amazonas. Zool. Mus. Berlin 21, pages 264-269.

-, 1936(b): Beschreibung dreier neuer Fische der Familie *Cichlidae* aus Südamerika. Sitzber. Ges. natf. Freunde Berlin, pages 138-142.

-, 1938: Über einen neuen südamerikanischen Fisch der Familie Cichlidae. Zool. Anz. 123, pages 246-247. (*A. aequipinnis* from Argentina).

Arnold, J.P. & Ahl, E., 1936: Fremdländische Süßwasserfische. Gustav Wenzel & Sohn, Braunschweig, 592 pages.

Böhme, T., Fey, S., Bork, D., & Bergk, S., 1990: Fischparadies Französisch-Guayana: Auf der Spur von *Apistogramma gossei*. Das Aquarium 24 (10/256), pages 13-16.

Bork, D., 1996: *Apistogramma mendezi* & Co. Seltene *Apistogramma*-Arten aus dem Einzugsgebiet des Rio Negro und Rio Uaupés. Das Aquarium 30 (6/324), pages 10-14.

-, D., 1996: *Apistogramma* sp. "Pandurini" - Eine schöne Bescherung aus Peru. Das Aquarium, 30 (10/328), pages 23-26.

-, & Mayland, H.J., 1996: *Apistogramma linkei* - Ein Zwergcichlide aus Bolivien ohne züchterische Probleme. Das Aquarium, 30 (2/320), pages 22-23

Boulenger, G.A., 1895: Viaggo del dottor Alfredo Borelli nella Repubblica Argentina e nel Paraguay. XII. Poissons. Bol. Mus. Zool. Anat. Cp. R Univ. Torino 10 (196), pages 1-3.

-, 1900: Viaggio del Dr. A. Borelli nel Matto Grosso e nel Paraguay. III. Liste des poissons recueillis á Urucum et á Carandasinho, prés de Corumbá. Boll. Mus. Zool. Anat. Cp. R. Univ. Torino 15 (370), pp. 1-4.

Cope, E.D., 1872: On the fishes of the Ambyiacu River. Proc. Acad. Nat. sci. Philadelphia 1871, pp. 250-294.

Eigenmann, C. H., 1911: The localities at which Mr. John D. Haseman made collections. Ann. Carnegie Mus. 7, pp. 299-314.

-, 1912: The freshwater fishes of British Guiana, including a study of the ecological groupings of species, and the relation of the fauna of the plateau to that of the lowloands. Mem. Carnegie Mus. 5, pp. 1-578.

-& Bray, W.L., 1894: A revision of the American *Cichlidae*. Ann. New York Acad. Sci. 7, pp. 607-624.

-& Kennedy, C.H., 1903: On a collection of fishes from Paraguay, with a synopsis of the American genera of Cichlids. Proc. Acad. Nat. sci. Philidelphia 5, pp. 497-537.

-, McAtee, W.L. & Ward, D.P., 1907: On further collections of fishes from Paraguay. Ann. Carnegie Mus. 4, pp. 110-157.

Eschmeyer, W.N., 1990: Catalog of the Genera of recent fishes. California Academie of Sciences, San Francisco, 697 pp.

Fischer, P.H., 1968: *Apistogramma ramirezi* ist doch Venezuelaner! Datz 21 (1), pages 8-10.

Fowler, H.W., 1914: Fisches from the Rupununi River, British Guiana. Proc. Acad. Nat. Sci. Philadelphia 66, pp. 229-284.

-, 1954: Os peixis de água doce do Brazil, Vol. II; Arg. Zool. Est. Sao Paulo 9, pp. 1-400.

-, 1954: Os peixis de água doce do Brazil, Vol IX; Arg. Zool. Est. Sao Paulo 9, pp. 1-400.

Frey, H., 1957: Das Aquarium von A bis Z. Neumann-Verlag, Leipzig, 1. Edition.

Géry, J.,1983: Le nom de genre de *Apistogramma ramirezi* Myers & Harry. Revue fr. Aquariol. 10, (3), pp. 71-72.

Goldstein, R.J., 1970: Cichlid notes. T.F.H. 18 (12). Pp. 78-81.

-, 1973: Cichlids of the world. T.F.H. Publ. , Neptune City/ U.S.A., 382 pp.

Grad, C., 1986: Mein erster Zuchterfolg: *Apistogramma inconspicua*. DCG-Info 17 (10), pages 199-200.

Günther, A.C.L.G., 1862: Catalogue of the fishes of the British Museum, vol. IV, British Museum, London, 534 pp.

Haseman, J.D., 1911(a): A brief report upon the expedition of the Carnegie Museum to Central South America. Ann. Carnegie Mus. 7, pp. 287-299.

-, 1911(b): Description of some new species of fishes and miscellaneous notes on others obtained during the expedition of the Carnegie Museum to Central South America. Ann. Carnegie Mus. 7, pp. 315-328.

-, 1911 (c): An annotated catalog of the cichlid fishes collected by the expedition of the Carnegie Museum to Central South America, 1907-10. Ann. Carnegie Mus. Ser. 7, pp. 329-373.

Heymer, A., 1977: Ethologisches Wörterbuch, 1. Edit., Parey, Berlin/Hamburg, 237 pages.

HOEDEMAN, J.J., 1951: Notes on the fishes of the cichlid family. I. *Apistogramma cacatuoides* sp. n. Beaufortia (4), pp. 1-4.

HOLLY, M., MEINKEN, H. & RACHOW, A., 1927: (Lose leaf edition): Die Aquarienfische in Wort und Bild. Kernen-Verlag, Stuttgart.

ISBRÜCKER, I.J.H., 1986: *Microgeophagus* FREY, 1957 (synonym *Papiliochromis* KULLANDER, 1977). Het Aquarium 56 (12), pp. 312-313.

KANAZAWA, R.H., 1966: The fishes of the genus *Osteoglossum* with a description of a new species from the Rio Negro. Ichthylogica Aquar. J. 37, pp. 161-172.

KILIAN, B., 1992: Freilandbeobachtungen an *Dicrossus* . DATZ 45 (3), page 143.

KOSLOWSKI, I., 1981: Neues aus der *Apistogramma*-Szene. DCG-Info 12 (8), pages 141-144.

-, 1984: Neuheiten und Neuigkeiten aus der Gattung *Apistogramma*, Teil I. DCG-Info 15 (3), pages 50-55.

-, 1985(a): Die Buntbarsche der Neuen Welt, Zwergcichliden. Edition Kernen, Essen, 192 pages.

-, 1985(b): Die Balzfleck-*Apistogramma*, ein Kuriosum unter dem Zwergbuntbarschen Südamerikas. DATZ 38, pages 102-105.

-, 1994(a): *Apistogramma*-Arten aus dem Rio Tefé. DATZ 47 (3), pages 152-156.

-, 1994(b): Ostamazonische *Apistogramma*. DATZ 47 (12), pages 781-788.

-, 1995: Beiträge zur Unterscheidung von *Apistogramma*-Weibchen. In ›Cichliden‹, Festschrift zum 25jährigen Jubiläum der DCG, Frankfurt am Main, (256 p.), pp 204-217.

KULLANDER, S.O., 1976: *Apistogramma luelingi* sp. nov., a new cichlid fish from Bolivia. Bonn. Zool. Beitr. 27, pages 258-266.

-, 1978: A rediscription of *Crenicara filamentosa* LADIGES, 1958 (Teleostei: Cichlidae). Mitt. Hamb. Zool. Mus. Inst. 75, pages 266-278.

-, 1979: Species of *Apistogramma* (*Teleostei*, *Cichlidae*) from the Orinoco drainage basin, South America, with description of four new species. Zool. Scripta 8, pp. 69-79.

-, 1980(a): A taxonomic study of the genus *Apistogramma* REGAN, with a revision of Brazilian and Peruvian species (*Teleostei*: *Percoidei*: *Cichlidae*). Bronner Zool. Monograph. 14, 152 pages.

-, 1980(b): A rediscription of the South American Cichlid Fish *Papiliochromis ramirezi* (MYERS & HARRY). Stud. Neotrop. Fauna Envir. 15, pp. 91-108.

-, 1981(a): Description of a new species of *Apistogramma* (*Teleostei*: *Cichlidae*) from the upper Amazonas basin. Results from the Argentinien-Peruvian Expedition Dr. K. H. Lüling 1978. Bonner zool. Beitr. 32 (1-2) pages 183-194.

-, 1981(b): The Bolivian Ram, a zoogeographical problem and its taxonomic solution. DCG-Info, 12. Pages 61-79.

-, 1982(a): Beschreibung einer neuen *Apistogramma*-Art aus Zentral-Amazonien. DCG-Informationen 13, pages 181-193.

-, 1982(b): Description of a new species of *Apistogramma* REGAN, from the Oyapock and Approuague river systems. Cybium 6 (3), pp. 65-72.

-, 1986: Cichlid fishes of the Amazon River drainage of Peru. Swedish Museum of Natural History, Stockholm, 431 pp.

-, 1987: A new *Apistogramma* species (Teleostei: Cichlidae) from the Rio Negro in Brazil and Venezuela. Zool. Scr. 16 (3), pp. 259-270.

-, 1989: *Biotoecus* EIGENMANN & KENNEDY (*Teleostei*: *Cichlidae*): description of a new species from the Orinoco basin and revised generic diagnosis. J. Nat. Hist. 23, pp. 225-260.

-, 1990: *Mazarunia mazarunii* (*Teleostei*: *Cichlidae*), a new genus and species from Guyana, South America. Ichthyol. Explor. Freshwaters, Vol. I (1), pages 3-14.

- & NIJSSEN, H., 1989: The cichlids of Surinam. E.J. Brill, Leiden, The Netherlands, 256 pp.

KUSTERS, J., 1991: Begegnung mit *Apistogramma nijsseni*. DATZ 44, pages 386-388.

LINKE, H., 1987: Es ist gelungen: die Mehrung von *Papiliochromis altispinosa*. Aquarium heute 5 (1), pages 12-14.

- & STAECK, W., 1995/5. Ed.: Amerikanische Cichliden I: Kleine Buntbarsche. Tetra-Verlag, Melle, 232 pages.

- & STAECK, W., 1986: Neu im Aquarium: *Apistogramma luelingi* und *Papiliochromis altispinosa*. Aquarium heute 4 (1), pages 7-9.

LÜLING, K.H., 1969(a): Am Fundort des *Apistogramma ramirezi* in Bolivien. Das Aquarium, Wuppertal 3 (16), pages 114-117.

-, 1969(b): Auf Fischfang in den Urwäldern am Rio Chaparé und Rio Chipiriri in Ostbolivien. Aqua Terra 6 (5), pgs 56-60; (6), pgs 65-72; (7), pgs 73-81.

-, 1977: Zierfische vom Rio Paraná. Aquar. Mag. 11 (6), pages 221-224.

-, 1980: Das Fischparadies Yarina Cocha. Exkursion zum mittleren Ucayali in Ostperu. Aqu. Mag. 14 (5), pages 248-254.

-, 1981: Zwei unterschiedliche Fließwasserbiotope im Einzugsgebiet des mittleren Ucayali (Ostperu) und ihre Fische. Results from the Argentina-Peru-Expedition Dr. K.H. Lüling 1978. Bonn. Zool. Beit. 32, pages 167-182.

-, 1984: Die Fischwelt im Schatten der Aguaje-Palme in Westamazonien nebst kurzer Notiz ueber den Lebensbereich von Segelflossen und Flaggenbuntbarschen. Tetra-Info No.68/4, pages 5-9.

MAYLAND, H.J., 1978: Große Aquarienpraxis, Band III, Cichliden und Fischzucht. Landbuch-Verlag, Hannover, 461 pages.

-, 1980: Fischparadies Chaco. Aqua. Mag. 14, pp 128-135.

-, 1981/1992: Diskusfische - Könige Amazoniens. Landbuch-Verlag, Hannover, 240 pages.

-, 1987/1990: Diskusfibel. Landbuch-Verlag, Hannover, 208 pages.

-, 1987: Ein neuer Zwergcichlide der Gattung *Apistogramma*, (São-Gabriel-Apistogramma). DATZ 40, pages 345-347.

-, 1988: Diskusfieber. Landbuch-Verlag, Hannover, 215 pages.

-, 1994: Breitbinden-Zwergcichlide - Zur Fortpflanzung von *Apistogramma* sp. "Breitbinden". Das Aquarium, Bornheim, 28 (297/3), pages 15-16.

-, 1995(a): Praxis Diskusaquarium. Landbuch-Verlag, Hannover, 176 pages.

-, 1995(b): *Microgeophagus altispinosus* - eine neue Variante (?) aus dem brasilianischen Mato Grosso. Das Aquarium, Bornheim, 29 (318/12), pages 10-11.

-, 1995(c): Cichliden. Landbuch-Verlag, Hannover, 596 pages.

MAYR, E., 1975: Grundlagen der zoologischen Systematik (Translated by Prof. Dr. Otto Kraus). Verlag Paul Parey, Hamburg/Berlin, 370 pages.

MEINKEN, H., 1961: Drei neu eingeführte Arten aus Peru, eine davon wissenschaftlich neu (*A. sweglesi*). DATZ 14 (5), pages 135-139.

-, 1962: Eine neue *Apistogramma*-Art aus dem mittleren Amazonas-Gebiet, zugleich mit dem Versuch einer Ubersicht über die Gattung. Senckenberg. Biol. 43, pages 137-143.

-, 1964: *Apistogramma kleei* spec. Nov., der Querbinden-Zwergbarsch. Aquarien Terrarien 17, pages 293-297.

-, 1965(a): Über eine neue Gattung und Art aus Venezuela. Senckenberg. Biol. 46, pages 47-53.

-, 1965(b): Eine neue *Apistogramma*-Art aus Venezuela. Senckenberg. Biol. 46, pages 257-263.

-, 1967: Wiederum platzte eine Import-Legende. DATZ 20 (10), pages 294-296.

-, 1968: Nochmals: "*Apistogramma ramirezi* ist doch Venezuelaner". DATZ 21 (4), pages 107-109.

-, 1969(a): Zur Frage des Vorkommens von *Apistogramma ramirezi* MYERS & HARRY, 1948, Aquar. Terr. 16 (5), pages 165-166.

-, 1969(b): *Apistogramma gibbiceps* n. sp. Aus Brasilien. Senckenberg. Biol. 50, pages 91-96.

-, 1970: *Apistogramma gibbiceps*, a new species from Brazil. Buntbarsche Bull. (US-Amerik. Cichlid Assn) (24), pages 7-10.

-, 1971(a): Bekommt *Apistogramma ramirezi* MYERS & HARRY, 1948, einen anderen Gattungsnamen? DATZ 24 (7), pages 224-225.

-, 1971(b): *Apistogramma geisleri* n. sp. Und *Apistogramma borellii* (Regan) aus dem Amazonas-Becken. Senckenberg. biol. 52, pages 35-40.

MENDES, C., 1990: Rettet den Regenwald! Lamuv Verlag GmbH, Göttingen, in partnership with the Deutsche Welthungerhilfe, 157 pages.

MYERS, G.S., 1935: Four new fresh-water fishes from Brazil, Venezuela and Paraguay. Proc. Biol. Soc. Washington 48, pp. 1-14.

- & HARRY, R.R., 1948: *Apistogramma ramirezi*, a cichlid fish from Venezuela. Proc. California zool. Club 1 (1), pp. 1-8.

NIEUWENHUIZEN, VAN DEN, A., 1990: Papiphagus-Krimi. DATZ 43 (1), pages 51-53.

OHM, D., 1980: Ein Buntbarsch wechselt sein Geschlecht: Bei *Crenicara punctulata* werden Weibchen zu Männchen. DCG-Info 11 (9), pages 161-170.

-, 1980: Weibchen werden Männchen. Geschlechtsumwandlung beim Buntbarsch *Crenicara punctulata*. Aquarien Magazin 14 (11), pages 631-634.

PELLEGRIN; J., 1902: Cichlidés du Brésil rapportés par. M. Jobert. Bull. Mus. Hist. Nat. Paris 8, pp. 181-184.

-,1936: Un poisson d'aquarium nouveau du genre *Apistogramma*. Bull. Soc. Natl. Acclimat. Prot. Nat. 83, pages 56-58.

REGAN, C.T., 1905(a): A revision of the fishes of the South American Cichlid Genera *Crenacara*, *Batrachops*, and *Crenicichla*. Proc. Zool. Soc. London 1, pp. 152-168.

-, 1905(b): A Revision of the Fishes of the South American Cichlid Genera *Acara*, *Nannacara*, *Acaropsis*, and *Astronotus*. Ann. Gag. Nat. Hist. (7) 15, pp. 329-427.

-, 1906: A Revision of the South American Cichlid Genera *Retroculus*, *Geophagus*, *Heterogramma*, and *Biotoecus*. Ann. Mag. Nat. Hist. (7) 17, pp. 49-66.

-, 1913: Fishes from the River Ucayali, Peru, collected by Mr. Mounsey. Ann. Mag. Nat. Hist. 6 (8), pp. 281-283.

RAHM, P. DE & KULLANDER, S.O., 1983: *Apistogramma nijsseni* KULLANDER un nouveau Cichlidé nain pour l'aquarium. Revue fr. Aquariol. 9 (4/82), pp. 97-104.

RICHTER, H.J., 1987: Zwergbuntbarsche. Neumann Verlag Leipzig/Radebeul, 220 pages.

RÖMER, U., 1989: Zur Problematik der Geschlechterverteilung von Nachzuchttieren bei *Apistogramma nijsseni* KULLANDER, 1979. DCG-Info 20 (4) pages 74-77.

-, 1991(a): Zur Lebenserwartung von Zwergbuntbarschen der Gattung *Apistogramma*. DCG-Info 22 pages 70-72.

-, 1991(b): Der Smaragd-Zwergbuntbarsch. DATZ 44 (5), pages 290-292.

-, 1992: Beobachtungen zur Aquarienbiologie von *Apistogramma diplotaenia* KULLANDER, 1987. DCG-Info 23 (8), pages 166-170.

-, 1995: Warnung vor Zwergcichlidenhybriden: *Apistogramma* sp. "Mariae". DATZ 48 (1), pages 18-20.

-, 1995: Uaupés. Aqua geógraphia 4 (1), pages 6-27.

- & BEISENHERZ, W., 1995: Modifikatorische Geschlechtsbestimmung durch Temperatur und pH-Wert bei Buntbarschen der Gattung *Apistogramma*. Im Symposiumband: Fortpflanzungsbiologie der Aquarienfische, B. Schmettkamp Verlag, Bornheim, 271 pages (pgs. 261-266).

- & SOARES, D., 1996: *Apistogramma juruensis* KULLANDER, 1986: Beobachtungen zur Aquarienbiologie eines neu eingeführten Zwergbuntbursches. Aquarium heute (3), pages 356-359.

SCHMETTKAMP, W., 1976: Der Schönflossige Zwergbuntbarsch: *Apistogramma ornatipinnis* AHL, 1936 DCG-Info 7 (7), pages 121-126.

-, 1977: Bemerkungen zur Gattung *Apistogramma*, speziell zu *Apistogramma ornatipinnis*. Das Aquarium 11 (92), pages 59-63.

-, 1978: Neues aus der *Apistogramma*-Szene: Rotpunkt-, Wangenfleck-, Segelflossen- und Schwarzbinden-Apistogramma. DCG-Info 9 (1), pages 16-20.

-, 1981: Neues aus der *Apistogramma*-Szene. DCG-Info 12 (8), pages 148-150.

-, 1982: Die Zwergcichliden Südamerikas. Landbuch-Verlag, Hannover, 176 pages.

-, 1985: Der Wangenfleck-Zwergbuntbarsch: *Apistogramma resticulosa* KULLANDER, 1980. DCG-Info 16 (6), pages 116-119.

SIOLI, H., 1983: Amazonien - Grundlagen der Ökologie des größten tropischen Waldlandes. Wissenschaftl. Verlagsgesellschaft, Stuttgart, (Paperback d. Zeitschr. Naturwiss. Rundschau), 64 pages.

STAECK, W., 1985: Erfahrungen mit zwei neuen *Apistogramma* aus Bolivien; 1. Die Querstreifen-Apistogramma. DATZ 38 (1), pages 4-7.

-, 1985: Erfahrungen mit zwei neuen *Apistogramma* aus Bolivien; Die Gelbbrust-Apistogramma. DATZ 38 (2), pages 60-63.

-, 1986: Beiträge zur Kenntnis peruanischer Zwergcichliden: 2. *Apistogramma eunotus*, KULLANDER, 1981. DATZ 39 (12), pages 546-548.

-, 1987(a): *Papiliochromis altispinosa*. Ein "missing link"' in der Evolution der Maulbrutpflege? DCG-Info 18 (6), pages 115-117.

-, 1987(b): Beiträge zur Kenntnis peruanischer Zwergcichliden. 5. *Apistogramma agassizii* (STEINDACHNER, 1875). DATZ 40 (6), pages 299-302.

-, 1987 (c): Beiträge zur Kenntnis peruanischer Zwergcichliden. 6. *Apistogrammoides pucallpaensis* MEINKEN, 1965. DATZ 40 (10), pages 439-441.

-, 1990(a): Neuer Schachbrettcichlide. DATZ 43 (12), pages 711-712.

-, 1990(b): *Apistogramma*-Arten Venezuelas. 1. Ergänzungen zur Beschreibung von *Apistogramma hoignei* MEINKEN, 1965. DATZ 43 (7), pages 412-416.

-, 1990(c): *Apistogramma*-Arten Venezuelas. 2. *Apistogramma steindachneri*. DATZ 43 (12), pages 732-734.

-, 1991(a): *Apistogramma*-Arten Venezuelas. 3. *Apistogramma* spec. "Rio Caura". DATZ 44 (9), pages 234-237.

-, 1991(b): *Apistogramma*-Arten Venezuelas. 4. *Apistogramma hongsloi* KULLANDER, 1979.DATZ 44 (9), pages 572-574.

-, 1991(c): Eine neue *Apistogramma*-Art (*Teleostei*: *Cichlidae*) aus dem peruanischen Amazonasgebiet. Ichthyol. Explor. Freshwaters, Vol. 2, Nr. 2, pages 139-149.

-, 1992: *Apistogramma*-Arten Venezuelas. 5. *Apistogramma guttata*. 45 (9) pages 584-587.

-, 1993(a): Neuer Schmetterlingsbuntbarsch. DATZ 46 (4) page 212.

-, 1993(b): Zur Verbreitung und Ökologie von *Papiliochromis ramirezi*. DATZ 46 (4), pages 239-242.

-, 1993(c): *Apistogramma urteagai* KULLANDER, 1986. Ein aquaristisch neuer Zwergcichlide. Buntbarschjahrbuch 1994, pages 74-82.

STEINDACHNER, F., 1875: Beiträge zur Kenntnis der Chromiden des Amazonasstromes. Sitzber. K. Akad. Wiss. Wien 71, pages 61-137.

SUTTNER, R., 1987: Die "grauen Mäuse" (*Apistogramma resticulosa* KULLANDER, 1980) überraschten mit ihrer Brutpflege. DCG-Info 18 (1), pages 16-19.

TEMBROCK, G., 1978: Verhaltensbiologie unter besonderer Berücksichtigung der Physiologie des Verhaltens. Gustav Fischer Verlag, Jena, 224 pages.

UNTERGASSER, D., 1989: Krankheiten der Aquarienfische. Frankh'sche Verlagshandlung, W. Keller & Co., Stuttgart, 176 pages.

WACHTER, T., 1983: *Apistogramma ortmanni*. Tetra-Info (Heft 63/3), pages 11-12.

WARZEL, F., 1996: Ein neuer Schachtbrettcichlide aus dem Rio Tapajós. Das Cichlidenjahrbuch, Band 6 (Red. A. Konings), Cichlid Press, St. Leon-Rot, Germany.

WERNER, U., 1989: Zum Fischfang im Mato Grosso; Teil 4: Von Sinop zu den Goldgräbern am Peixoto. DCG-Info 20 (11), pages 210-221.

WICKLER, W., 1960: Über die systematische Stellung von ›*Apistogramma*‹ *ramirezi* MYERS & HARRY. Aquar. Terr. 7, Leipzig, Berlin, pages 327-328.

WINDISCH, W. , 1989: *Apistogramma* spec. Rußkopf. DCG-Info 20 (10), pages 199-201.

-, 1991: Neues von *Apistogramma diplotaenia*. DATZ 44, page 214.

-, 1991: Ein neuer Zwerg: Weißsaum-Apistogramma. DATZ 44, page 335.

-, 1993: *Apistogramma* cf. *agassizii* und seine Farbformen am Rio Negro. DCG-Info 24 (1), pages 14-19.

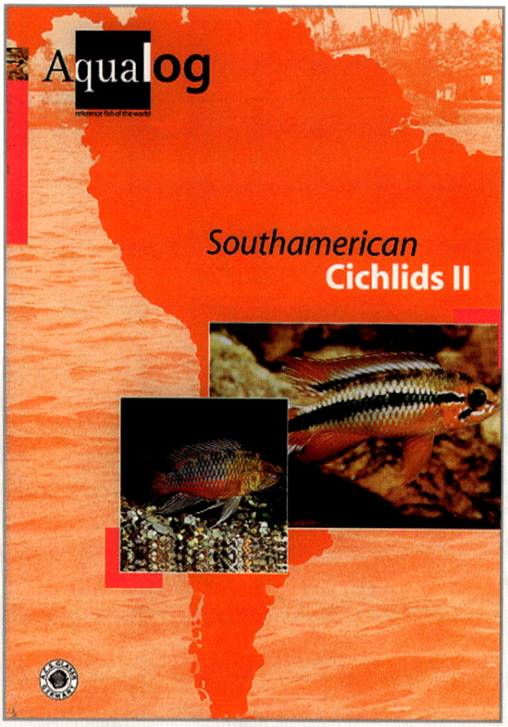

The unique reference work with the supplementing sticker-system

ISBN: 3-931702-07-3

This AQUALOG includes:
More than 500 full-colour photos
All scientifically known species of the genera Apistogramma, Nannacara, Biotoecus, Dicrossus, Crenicara, Microgeophagus, Taeniacara and Apistogrammoides in more than 220 different varieties
Plus identification keys, maps , care instructions ...
AQUALOG is the first reference catalogue offering a supplementing system for keeping your books up-to-date for years and years - with stickers of the latest discoveries

First Supplement for Southamercian Cichlids II

Apistogramma sp. "Pandurini",
Apistogramma sp. "Paraguay I"
Apistogramma sp. "Paraguay II"
Apistogramma cf. brevis II
Apistogramma pertensis "Peru-Whiteseam"
Apistogramma piauiensis "Red-Blue"

ISBN: 3-931702-12-X

ISBN: 3-931702-01-4;
112 pages, 483 photos
All Suckermouth Cats with L- and LDA-numbers L001-L202 and LDA 01 - LDA 25 (with all stickers up to L238 and LDA 30)

ISBN: 3-931702-04-9;
112 pages, 466 photos
All Cichlids of the genera *Cichla, Crenicichla, Geophagus, Guianacara, Gymnogeophagus, Retroculus, Satanoperca, Teleocichla* and *Uaru*

ISBN: 3-931702-07-3;
112 pages, 466 photos
All Cichlids of the genera *Apistogramma, Apistogrammoides, Biotoecus, Crenicara, Dicrossus, Microgeophagus, Nannacara* and *Taeniacara*

ISBN: 3-931702-10-3;
144 pages, 673 photos
All Cichlids of the former genera *Cichlasoma* and *Aequidens*

ISBN: 3-931702-13-8;
144 pages, 674 photos
All Plated Cats of the genera *Aspidoras, Brochis, Corydoras, Dianema, Hoplosternum, (Lepthoplosternum, Megalechis)* and *Callichthys*

ISBN: 3-931702-21-9;
144 pages, 616 photos
All Anabantoids, Nandids, Snakeheads of the genera *Anabas, Belontia, Betta, Colisa, Ctenopoma, Ctenops, Helostoma, Luciocephalus, Macropodus, Malpulutta, Osphronemus, Parasphaerichthys, Parosphromenus, Pseudosphromenus, Sandelia, Sphaerichthys, Trichogaster, Trichopsis, Badis, Pristolepis, Afronandus, Nandus, Monocirrhus, Polycentrus, Polycentropsis* and *Channa*

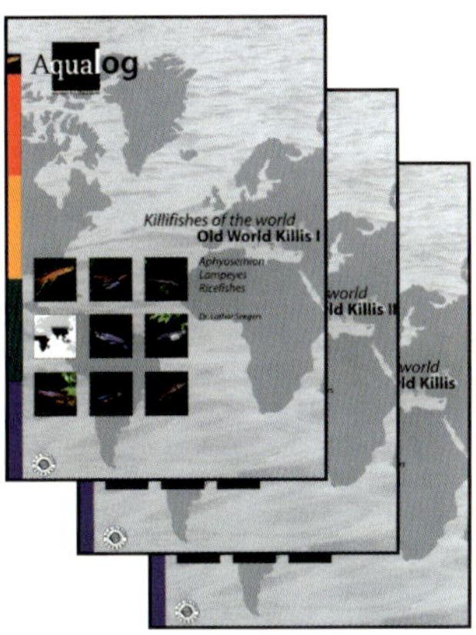

ISBN: 3-931702-25-1;
160 pages, 907 photos
All Ricefishes, Lampeyes and Killis of the genera *Adamas, Aplocheilichthys, Aphyoplatys, Aphyosemion, Laciris, Hylopanchax, Lamprichthys, Pantanodon, Procatopus, Oryzias, Xenopoecilus, Adrianichthys* and *Horaichthys*

In preparation for 1997

Killifishes of the World - Old World Killis II
(with *Aphanius, Nothobranchius, Aplocheilus, Epiplatys* etc.)
(November 1997)

Southamerican Cichlids IV
(with all Discus and Angelfishes, including breeding forms)

Verlag A.C.S., Rothwiesenring 5,
D-64546 Mörfelden-Walldorf,
fax: (0049) (0) 6106 - 644692
tel.: (0049) (0) 6106 - 690140

The complete, highly actual reference work you cannot do without!

The renowned AQUALOG system - used worldwide by importers & expor

- AQUALOG Specials - presenting the most popular and beautiful fishes of one group
- Each special makes a beautiful fold-up poster
- Includes an informative introduction and care instructions

special;

The complete reference work

- Get your subscription for the one and only newspaper for hobby-ists
- Multi-coloured and full of the latest news and information on fishes and aquarium keeping
- Includes Stickups of the newest discoveries and breeding forms for keeping your AQUA-LOG book up-to-date

news;

For more information visit our homepage: http://www.aqualog.de. Don´t hesitate to contact us! e-mail: acs@nacamar.de

Supplements published so far

Supplement No.1 for
Loricariidae all L-Numbers
ISBN 3-931702-15-4
L203, L204, L205, L30a, L200

Supplement No.2 for
Loricariidae all L-Numbers
ISBN 3-931702-16-2
L200, L206, L207, L192

Supplement No.3 for
Loricariidae all L-Numbers
ISBN 3-931702-17-0
L208, L209, L210, L211, LDA 26

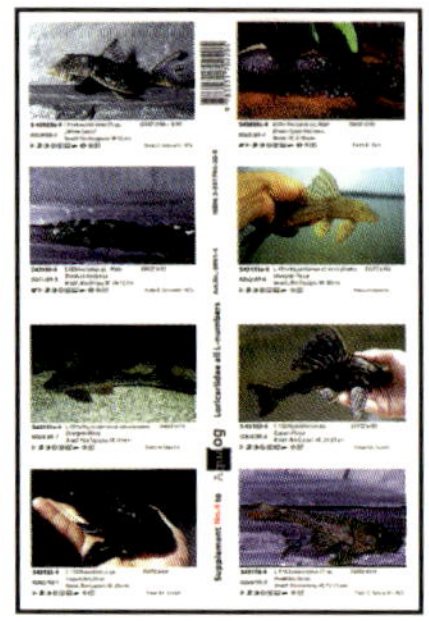

Supplement No.4 for
Loricariidae all L-Numbers
ISBN 3-931702-20-0
L33a, L59c, L100, L131a, L132, L133, L178

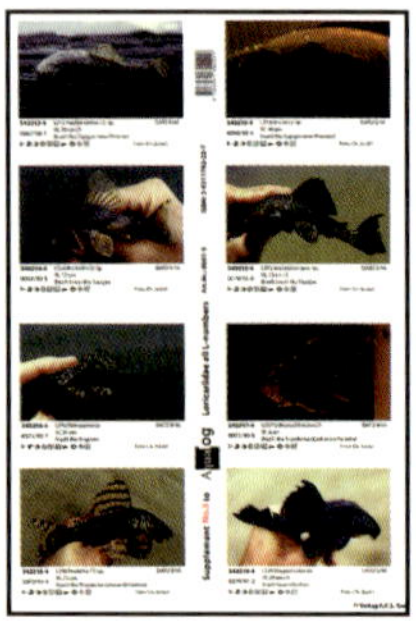

Supplement No.5 for
Loricariidae all L-Numbers
ISBN 3-931702-22-7
L212, L213, L214, L215, L216, L217, L218, L219

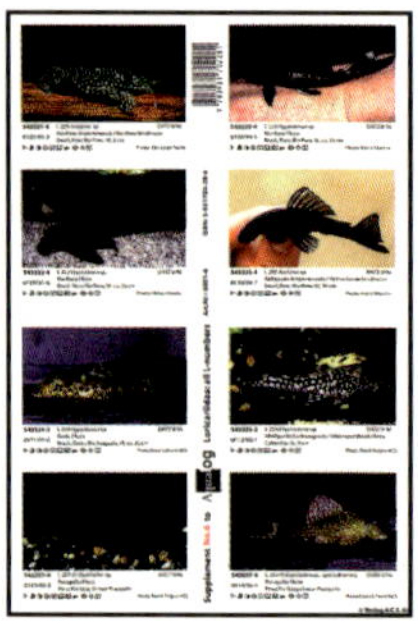

Supplement No.6 for
Loricariidae all L-Numbers
ISBN 3-931702-28-6
L221, L222, L223, L224, L225, L227

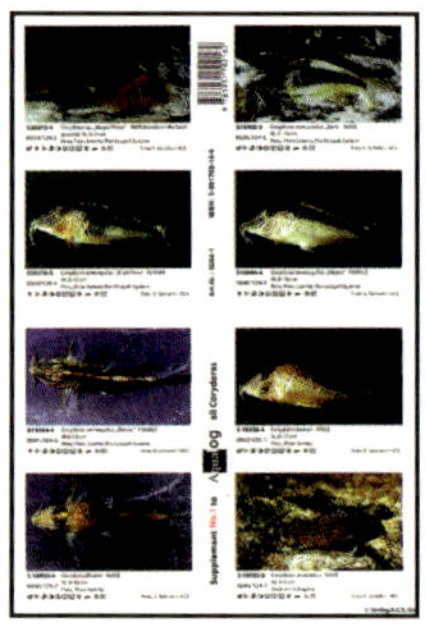

Supplement No.1 for
all Corydoras
ISBN 3-931702-18-9
C52, Corydoras semiaquilus, C. fowleri, C. punctatus

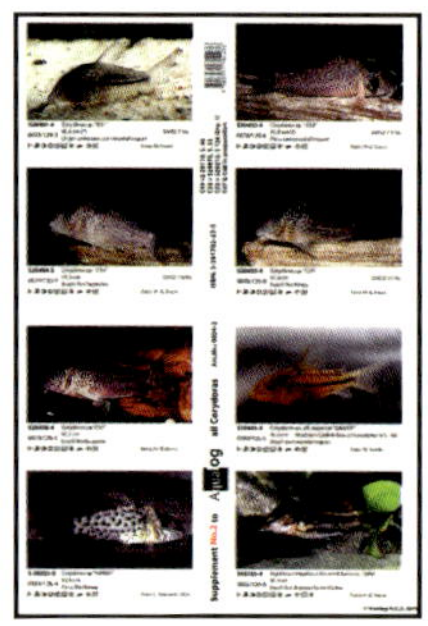

Supplement No.2 for
all Corydoras
ISBN 3-931702-23-5
C51, C53, C54, C55, C56, C. sp. aff. nattereri, C. sp. "Nanay", Aspidoras virgulatus

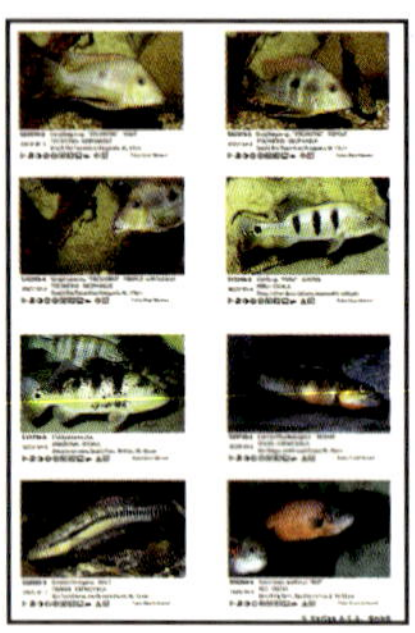

Supplement No.1 for
Southamerican Cichlids I
ISBN 3-931702-19-7
Geophagus sp. "Toncantins", Cichla sp. "Peru", C. monoculus, Crenicichla phaiospilus, C. tigrina, Astronotus ocellatus "Red"

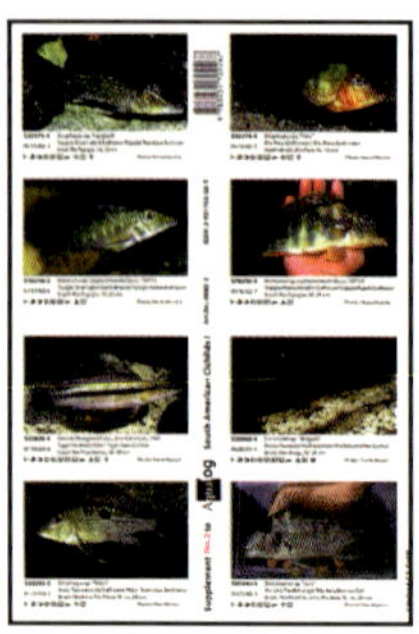

Supplement No.2 for
Southamerican Cichlids I
ISBN 3-931702-26-X
Geophagus sp. "Tapajos III", G. sp. "Paru", G. sp. "Maici", Retroculus sp., Crenicichla tigrina, C. sp. Xingu III, Satanoperca sp. "Jaru"

Supplement No.1 for
Southamerican Cichlids II
ISBN 3-931702-12-X
Apistogramma sp. "Pandurini", A. sp. "Paraguay I & II", A. cf. brevis II, A. pertensis, A. piauiensis

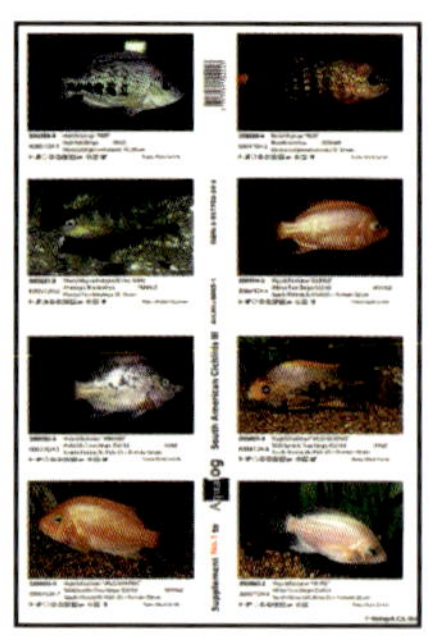

Supplement No.1 for
Southamerican Cichlids III
ISBN 3-931702-24-3
Herichthys sp. "Red", Thorichthys callolepis, Vieja bifasciatus (4 variaties)